MEMOIRS

of the
American Mathematical Society

WITHDRAWN

Number 446

A Sufficient Criterion for a Cone to be Area-minimizing

Gary R. Lawlor

May 1991 • Volume 91 • Number 446 (third of 4 numbers) • ISSN 0065-9266

American Mathematical Society
Providence, Rhode Island

1980 *Mathematics Subject Classification* (1985 *Revision*).
Primary 49F22; Secondary 49F10.

Library of Congress Cataloging-in-Publication Data

Lawlor, Gary R. (Gary Reid), 1961–
  A sufficient criterion for a cone to be area-minimizing/Gary R. Lawlor
  p. cm. – (Memoirs of the American Mathematical Society, ISSN 0065-9266; no. 446)
  "May 1991, volume 91, number 446 (third of 4 numbers)."
  Includes bibliographical references.
  ISBN 0-8218-2512-7
  1. Geometric measure theory. 2. Cone. I. Title. II. Series.
QA312.L34 1991
510 s–dc20   91-8060
[515´.42]   CIP

Subscriptions and orders for publications of the American Mathematical Society should be addressed to American Mathematical Society, Box 1571, Annex Station, Providence, RI 02901-1571. *All orders must be accompanied by payment.* Other correspondence should be addressed to Box 6248, Providence, RI 02940-6248.

SUBSCRIPTION INFORMATION. The 1991 subscription begins with Number 438 and consists of six mailings, each containing one or more numbers. Subscription prices for 1991 are $270 list, $216 institutional member. A late charge of 10% of the subscription price will be imposed on orders received from nonmembers after January 1 of the subscription year. Subscribers outside the United States and India must pay a postage surcharge of $25; subscribers in India must pay a postage surcharge of $43. Expedited delivery to destinations in North America $30; elsewhere $82. Each number may be ordered separately; *please specify number* when ordering an individual number. For prices and titles of recently released numbers, see the New Publications sections of the NOTICES of the American Mathematical Society.

BACK NUMBER INFORMATION. For back issues see the AMS Catalogue of Publications.

MEMOIRS of the American Mathematical Society (ISSN 0065-9266) is published bimonthly (each volume consisting usually of more than one number) by the American Mathematical Society at 201 Charles Street, Providence, Rhode Island 02904-2213. Second Class postage paid at Providence, Rhode Island 02940-6248. Postmaster: Send address changes to Memoirs of the American Mathematical Society, American Mathematical Society, Box 6248, Providence, RI 02940-6248.

Table of Contents

Contents

Abstract

A compact, k-dimensional surface (with boundary) is called area-minimizing if no other surface with the same boundary has less surface area.

An area-minimizing surface can have singularities. The main purpose of this paper is to investigate some of the shapes that such singularities can have. A key concept in this study is that of an area-minimizing cone.

We present a general method for proving that a cone with an isolated singularity is area-minimizing. The calculation involves the curvature (second fundamental form) and a sort of "embedding radius" of the normal bundle to the cone. We can also prove that certain cones are not area-minimizing.

Using this method, we complete the classification of minimizing cones over products of spheres. We also give other examples, including a family of unorientable minimizing cones.

The method also lends itself to perturbation arguments. We show that certain surfaces are area-minimizing in a small neighborhood of an isolated singularity.

Keywords and Phrases: geometric measure theory, area-minimizing surface, prescribed boundary problem, singular structure, cone

Received by editors January 24, 1989.

v

Acknowledgements

This Ph.D. thesis was done as a graduate student at M.I.T. and at Stanford. I wish to express thanks to those schools, and to the National Science Foundation for their support by way of a graduate fellowship.

I would like to thank my first advisor, Frank Morgan. I will always be grateful for his lessons of life and of mathematics.

Many thanks also to my advisor at Stanford, Brian White, who has been ever-available and very helpful in building my understanding and completing this work.

Others deserving grateful mention here are Robert Bryant and Leon Simon, for their help with Chapters 3 and 6, and the referee, for his prompt and helpful service.

Introduction

One of the fundamental objects of study in geometric measure theory is an "area-minimizing surface." By examining cones and small perturbations, this paper develops a sufficient criterion for a singular minimal surface to be area-minimizing when restricted to a small neighborhood of an isolated singularity.

Area-minimizing surfaces

A compact, k-dimensional surface (with boundary) is called area-minimizing if no other surface with the same boundary has less surface area. Soap films often model such a surface (for $k = 2$), since the surface tension tends to minimize area.

One immediate requirement for a surface to be area-minimizing is that it be minimal. A minimal surface is one whose curvatures balance each other out, so that a small deformation of the surface (holding the boundary fixed) does not change its area, to first order. If a surface is not minimal, then there is some direction in which we can deform a small piece of the surface and decrease the area without changing the boundary.

If a surface S is minimal, then a small piece of S cut out from a neighborhood of any regular point of S, is area-minimizing (see [M3, Cor. 3.4] and [F2]). In some cases, this is also true of small neighborhoods around singular points of S (see Theorem 6.3.1 herein).

Existence of area-minimizing surfaces

Given an arbitrary boundary, the first question we might ask is whether there exists an area-minimizing surface with that boundary. A fundamental theorem of geometric measure theory states that there is always an area-minimizing surface, as long as we adopt an appropriate definition of "surface." Manifolds won't serve the purpose; for some boundaries, there are sequences of manifolds with less and less area converging to an area-minimizing surface with singularities. The surfaces we will use are called "integral currents" (see [M5, section 3.10ff and ch. 4]). Roughly, an integral current can be thought of as the image of a manifold under a Lipschitz map, or a countable union of such images. This allows, for example, surfaces which are smooth everywhere except at singularities such as sharp points, creases, and self-intersections. It also allows objects with huge singular sets, which are hardly recognizable as "surfaces" in the intuitive sense.

1

Singularities

As mentioned above, some area-minimizing surfaces do have singularities. One of the main thrusts in recent work has been to understand the singularities that can occur in such surfaces. It is reassuring that given a fixed, smooth boundary, the surface (integral current) with that boundary having smallest possible area does not turn out to be one of these terrible, unrecognizable currents, but rather a nice, intuitive "surface," in general. For example, a two-dimensional area-minimizing surface sitting in $\mathbf{R}^3$ (with smooth boundary) has no singularities at all; it is always a manifold-with-boundary. In higher dimensions we begin to see singularities. A recent theorem of Almgren [A] tells us that the dimension of the singular set is at most two less than the dimension of the surface. A two-dimensional area-minimizing surface in $\mathbf{R}^4$, for example, has a (possibly empty) zero-dimensional singular set. Sheldon Chang has proved that the singular set in this case is a collection of isolated branch points and self-intersection points (see [ChS]).

Area-minimizing cones

The main purpose of this paper is to investigate some of the "shapes" that singularities in area-minimizing surfaces can take. A key concept in this investigation is that of an "area-minimizing cone." If B is a submanifold (or other subset) of the unit sphere, then the "cone over B" is the union of rays from the origin which pass through points of B. A cone is called area-minimizing if the "truncated cone" inside the unit ball is minimizing among all surfaces with boundary B.

A "tangent cone" to a surface S at a point $p \in S$ can be thought of as the union of rays extending from p and tangent to S at p. (See [M5, 9.7] for a more precise definition). This is a generalization of the notion of tangent plane. If the tangent cone at p is not a plane, then p is a singularity of S. Now if S is area-minimizing, then each tangent cone to S is also area-minimizing. Thus, in order to understand the first-order behavior (or "shape") of singularities in minimizing surfaces, we need to know which cones are area-minimizing.

Perhaps the simplest cone, besides a k-plane, is a pair of intersecting k-planes. This is the tangent cone we get at a self-intersection point of an immersed manifold. Frank Morgan's "Angle Conjecture," a beautifully simple condition telling which pairs of oriented k-planes are area-minimizing, was answered in the affirmative by complementary work of Dana Nance [N] and the author [L1].

This paper presents a sufficient condition for a cone to be area-minimizing. In principle it can be applied to the cone over any submanifold of the unit sphere, not requiring symmetry or orientability. The calculation involves the curvature (second fundamental form) and a sort of "embedding radius" of the normal bundle to the cone. Using this method, we complete the classification of minimizing cones over products of spheres. Another example given in the paper is the first known unorientable minimizing cone.

Historical notes

Benny Cheng has written an overview of the history of area-minimizing cones, and the methods of proving that they are minimizing [ChB]. The following is a brief history of minimization results for cones over products of spheres:

1969: Bombieri, DeGiorgi, and Giusti found the first codimension 1 minimizing singularities, which are $0 \ast (\mathbf{S}^k \times \mathbf{S}^k)$, for $k \geq 3$ (see [BDG]).

1972: Lawson added $0 \ast (\mathbf{S}^k \times \mathbf{S}^l)$ for $k + l \geq 7$ (see [Ls])

1974: Federer proved general conditions under which a singular differential form can be used as a calibration. Using singular calibrations, he obtained the same results as Bombieri, di Giorgi, and Giusti (see [F3, p. 405-407]). The calibrations of this paper can be viewed as a generalization of Federer's calibrations of cones over products of spheres, which eliminate the need for symmetry or special structure.

1974: P. Simoes added $0 \ast (\mathbf{S}^2 \times \mathbf{S}^4)$, and showed that $0 \ast (\mathbf{S}^1 \times \mathbf{S}^5)$ is not area-minimizing (see [Sm]). Incidentally, if C_1 is the part of the cone $0 \times (\mathbf{S}^1 \times \mathbf{S}^5)$ inside the unit ball, and S is the area-minimizing surface with boundary $\mathbf{S}^1 \times \mathbf{S}^5$, then the area of S is approximately

$$0.99999998 \cdot \mathrm{Area}(C_1).$$

1978: D. Bindshadler [B] found minimizing cones over products of m n-spheres, for $m = 3$, $n \geq 3$, or $m = 4$, $n \geq 5$, or $m \geq 5$, $n \geq m - 1$.

1987: Benny Cheng used methods from his thesis (M.I.T., 1987) to prove that $0 \ast (\mathbf{S}^k \times \mathbf{S}^l \times \cdots \times \mathbf{S}^l)$ is minimizing, for $k \geq 4$, $l \geq 4$. (Unpublished).

1987: Classification completed (Theorem 5.1.1).

Chapter 1: A Minimization Test for Cones

In this chapter we develop a method for proving that certain minimal cones are area-minimizing. The criterion we obtain can be applied by direct calculation. It depends on the curvature and dimension, and on the angles between certain tangent planes to the cone.

Section 1.1: The retraction Π

The main idea is to define a continuous, area-nonincreasing map $\Pi : \mathbf{R}^n \to C$ which holds C fixed. Then if C_1 is the "truncated cone" (the part of C inside the unit ball) and S has the same boundary as C_1, it will follow that

$$Area(S) \geq Area(\Pi(S)) \geq Area(C_1),$$

since $\Pi(S)$ must cover all of C_1.

Notation

Throughout this paper, we will let B denote a $k-1$ dimensional submanifold of the unit sphere in $\mathbf{R}^n$, and C the k-dimensional "cone over B," which is the union of rays extending from the origin and passing through B. For example, if B is a great circle of a 2-sphere, then C is a 2-plane. If C is not a k-plane, then it has an isolated singularity at the origin. The "truncated cone" C_1 is the part of C inside the unit ball.

In order to talk about area-minimizing surfaces, we need to first define our class of surfaces and boundaries, and our notion of area. In this paper, our orientable surfaces are integral currents, and area is defined as "mass," which is Hausdorff measure, counting multiplicity. For unorientable surfaces, we use integral currents modulo 2. Frank Morgan's book [M5] contains a beautifully clear development of currents, mass, area-minimization, and related concepts. See also [M6] and [F1, 4.2.26] for integral currents modulo ν.

General description

We now describe how to define a continuous retraction $\Pi : \mathbf{R}^n \to C$ which does not increase the area of any k-dimensional surface element.

Except for a radial shift, Π is just the "nearest point projection" onto C. That is, if z is a point of $\mathbf{R}^n$ and q is the point in the cone which is nearest to z, then

$\Pi(z)$ will be somewhere on the line segment $\overline{0q}$. The further z is from q, the closer $\Pi(z)$ will be to the origin; in particular, if the nearest point on C is not well-defined at z, then $\Pi(z) = 0$.

Since C is a union of rays from the origin, we can simplify the description of Π by restricting our attention to a single ray of C, and examining the subset of $\mathbf{R}^n$ which Π maps to the ray. If we delete the endpoint of the ray (the origin), then the inverse image will turn out to be a certain wedge-shaped, $n - k + 1$ dimensional set, which we will call a "normal wedge." To be precise, let us make the following definitions:

1.1.1 Definition: Let B be a smooth $k - 1$ dimensional submanifold of $\mathbf{S}^{n-1}$, and let $p \in B$. Define a "normal geodesic" of length α to be a unit speed geodesic (i.e., an arc of a great circle) σ which is orthogonal to B at its starting point $\sigma(0) = p$. If we leave off the endpoint $\sigma(\alpha)$, we will call σ an "open normal geodesic."

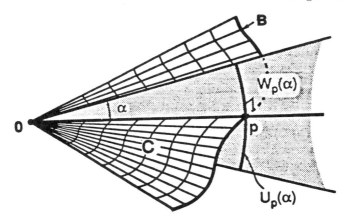

1.1.2 Definition of Normal Wedge: Let $B \subset \mathbf{S}^{n-1}$ as above, and let $C = 0 \times B$. For $p \in B$ and $\alpha > 0$, let $U_p(\alpha)$ be the union of all open normal geodesics of length α, and define the normal wedge $W_p(\alpha)$ as the cone $0 \divideontimes U_p(\alpha)$. For convenience, we leave out the origin, so that we can talk about "nonintersecting normal wedges."

For small α, the union of normal wedges over all $p \in B$ is a "conical neighborhood" of C in $\mathbf{R}^n$. For large α, the union covers $\mathbf{R}^n$. We will be interested in the cases where the normal wedges do not intersect each other.

The retraction we will define sends each W_p onto the ray $\overrightarrow{0p}$, along curves of projection similar to those shown in the following figure. That is, each curve is projected to its intersection with the ray. The boundaries of the wedges, and the

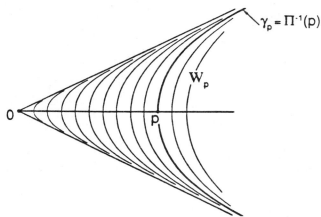

rest of $\mathbf{R}^n$ which is not covered by any wedge, are all projected to the origin. The projection curves in the normal wedge W_p are all determined by the single curve γ_p passing through p, by dilation from the origin. Thus, the retraction Π is "homothetically invariant," i.e., $\Pi(tz) = t\Pi(z)$.

The above figure, and the term "wedge," are suggested by the case when C is codimension 1, i.e., $k = n - 1$. In this case, there are only two normal directions to C at p. In higher codimension we extend the same projection curves in all of the

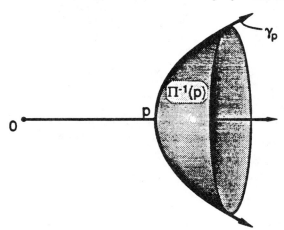

normal directions, so that we now obtain rotated surfaces of projection.

The curves γ_p, and thus the angular radius α of the wedges, will be determined as a function of the curvature and dimension of the cone C, in such a way as to make Π be area-nonincreasing. The angular radius of the wedges increases with greater curvature, and decreases with higher dimension of C, but is independent of the codimension $n - k$. If the cone is not minimal or if the curvature is too large for the given dimension k, an area-nonincreasing retraction will not exist. Also, even if the curvature is small enough so that the retraction can be defined on each wedge, the cone may be too "tightly packed" into $\mathbf{R}^n$, so that the wedges intersect each other, making the global retraction not well-defined. An example of this type of cone is a pair of planes through the origin which are very close to each other. The "normal radius" is a measure of how tightly packed the cone is:

1.1.3 Definition: The "normal radius" of a cone C at a point $p \in C$ is the angular radius α of the largest normal wedge $W_p(\alpha)$ that intersects C only in the ray $\overrightarrow{0p}$.

1.1.4 Example: In the 1-dimensional cone below, the normal radius at p is θ,

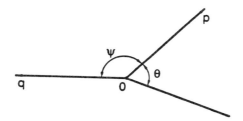

and the normal radius at q is ψ.

We summarize the above description of Π in the following definition:

1.1.5 Definition of the retraction Π: Suppose that for each point $p \in B$ we have specified a planar curve γ_p in terms of polar coordinates, such that $r(0) = p$, and $r(\theta) \to \infty$ as θ approaches some finite value $\theta_0(p)$. Let S_p be the curve or surface of revolution passing through p, consisting of the curve γ_p radiating in all directions normal to C, that is, sitting in planes containing the vector $\overrightarrow{0p}$ and a vector normal to C at p. Within the normal wedge $W_p(\theta_0(p))$ define a retraction Π_p by letting it send all of S_p to the point p, and be homothetically invariant, i.e., $\Pi_p(tz) = t\Pi_p(z)$, for $t \geq 0$, $z \in S_p$. Then if the normal wedges of radius $\theta_0(p)$ do not intersect, we define the global retraction $\Pi : \mathbf{R}^n \to C$ as equaling Π_p on W_p, and $\mathbf{0}$ outside the union of the normal wedges.

Derivation of γ_p

What remains is to derive the curves γ_p, in such a way that Π will not increase the area of any k-dimensional piece of surface; γ_p will be given in terms of a first order ordinary differential equation.

Zero curvature

We will simplify the discussion to begin with, by letting C be a k-plane, so that there is no curvature. For concreteness, let C be defined by the equations $x_{k+1} = \cdots = x_n = 0$, and let $p = (1, 0, \ldots, 0)$.

Note: To make the following argument precise, we take the limit as ϵ goes to zero; one way to do this would be to use the coarea formula (see [M5, 3.8]). For the sake of clarity, we'll suppress the limit argument and think of ϵ as being very small, or infinitesimal.

Consider the k-dimensional ϵ cube $E \subset C$, with corners at

$$p, \quad p(1 + \epsilon), \quad p + \epsilon v^1, \ldots, \quad p + \epsilon v^{k-1},$$

where v^i are unit vectors tangent to C, orthogonal to each other and to the vector $\overrightarrow{0p}$. We want to ensure that nothing with area (k-volume) less than ϵ^k projects onto

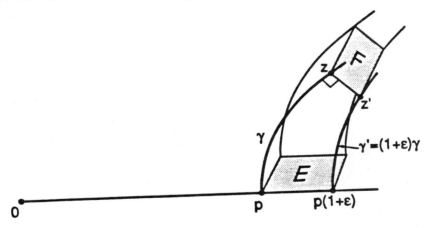

all of E. At a point $z \in \gamma$, the smallest k-dimensional element F that projects onto E will be orthogonal to the surfaces of projection. That is, F is spanned by the following vectors (with base point z): $\epsilon x \, v^1, \ldots, \epsilon x \, v^{k-1}$, and $z' - z$, where z' is the nearest point to z in the projection curve γ' through $p(1 + \epsilon)$, and where

$x = x_1$ is the horizontal distance from z to the origin, i.e., the component of z in the $\overrightarrow{0p}$ direction. Then $\mathrm{Vol}(F) = (\epsilon x)^{k-1}|z - z'|$. Thus, as we move out along γ, F is growing in $k-1$ directions, due to its increasing distance from the origin. This growth must offset the shrinking distance between the curves γ and γ'. In particular, we must require that

$$|z - z'| \geq \frac{\epsilon}{x^{k-1}}.$$

The larger the dimension k, the easier this will be to satisfy, even if the cone has some curvature; this helps to illustrate why there are more area-minimizing cones in higher dimensions.

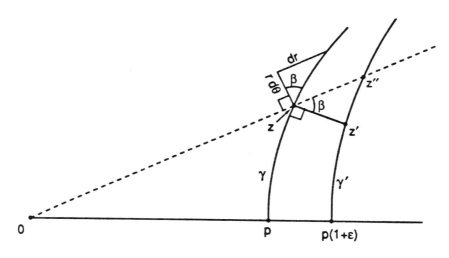

In the figure above, let $z = (r, \theta)$. Since the two curves γ and γ' are related by dilation from the origin by a factor of $1 + \epsilon$, we have $z'' = ((1 + \epsilon)r, \theta)$, so that

$$|z - z'| = r\epsilon \cos \beta = r\epsilon \cos\left(\tan^{-1}\frac{dr}{r\,d\theta}\right) = \frac{r^2\epsilon}{\sqrt{(dr/d\theta)^2 + r^2}}.$$

Then the requirement $|z - z'| \geq \epsilon/x^{k-1}$ becomes

$$\frac{r^2}{\sqrt{(dr/d\theta)^2 + r^2}} \geq \frac{1}{x^{k-1}} = \frac{1}{(r\cos\theta)^{k-1}}$$

$$r^{2k+2}(\cos\theta)^{2k-2} \geq \left(\frac{dr}{d\theta}\right)^2 + r^2$$

$$\frac{dr}{d\theta} \le r\sqrt{r^{2k}(\cos\theta)^{2k-2} - 1}, \ r(0) = 1.$$

As θ approaches some finite limit θ_0, $r \to \infty$. We wish to make θ_0 as small as possible; thus, we replace the inequality with equality. For $k \ge 3$, there is a one-parameter family of solutions to the differential equation (see Chapter 3). If we were to use the solution $r = \sec(\theta)$ (a straight line), then $\theta_0 = \pi/2$, and the resulting retraction would just be orthogonal projection onto the plane. The solution of interest is the one that goes to infinity at the smallest possible angle θ_0; for this solution, the angle θ, which we approximate numerically, is less than $\pi/2$. Its value in low dimensions is given in Table 1.4.1 (pp. 20-21) in the row corresponding to zero curvature.

Nonzero Curvature

Now if C has curvature, then Vol(F) gets smaller. Since there is still no curvature in the radial direction, the formula for $|z - z'|$ remains the same. However, the lengths of the other $k - 1$ vectors spanning F have changed. Since F is still orthogonal to the surface of projection, we can write

$$F = c(z - z') \wedge v^1 \wedge \cdots \wedge v^{k-1},$$

where c is some constant, and $\{v^i\}$ form an orthonormal basis for the tangent space to B at p. Further, since the expression we will derive below is independent of the choice of this basis, we can assume that each v_i is an eigenvector (with eigenvalue a_i) of the nearest-point projection π_N to C.

To be more precise, instead of using ϵ and suppressing the limit argument, let us now use $D\Pi$ acting on unit vectors.

Let $q = \pi_N(z) = (r\cos\theta)p$, so that we can write $z = q + t\nu$ for some unit vector ν normal to C at q. Replace $c(z - z')$ by a unit vector y lying in the plane spanned by $\frac{\partial}{\partial r}$ and ν, which is orthogonal to γ_p. Then

$$F = y \wedge v^1 \wedge \cdots \wedge v^{k-1}.$$

Our goal is to ensure that $\|E\| \le 1$, where $E = D\Pi(F)$. We have

$$D\Pi(F) = D\Pi(y) \wedge D\Pi(v^1) \wedge \cdots \wedge D\Pi(v^{k-1}).$$

We already worked out that

$$D\Pi(y) = \frac{\sqrt{(dr/d\theta)^2 + r^2}}{r^2} \frac{\partial}{\partial r}.$$

To compute $D\Pi(v^i)$, think of Π as being π_N followed by a (variable) shift toward the origin. Thus, for some set of constants $\{b_i\}$, we have

$$D\Pi : v^i \to (a_i v^i)\big|_q \to \left(\frac{a_i v^i + b_i \frac{\partial}{\partial r}}{|q|} \right) \bigg|_p$$

$$= \frac{a_i v^i + b_i \frac{\partial}{\partial r}}{r \cos \theta}.$$

Then letting

$$M = \left(\frac{\sqrt{(dr/d\theta)^2 + r^2}}{r^2} \right) \left(\frac{1}{(r \cos \theta)^{k-1}} \right),$$

we have

$$D\Pi(F) = M \left(\frac{\partial}{\partial r} \wedge (a_1 v^1 + b_1 \frac{\partial}{\partial r}) \wedge \cdots \wedge (a_{k-1} v^{k-1} + b_{k-1} \frac{\partial}{\partial r}) \right)$$

$$= M \frac{\partial}{\partial r} \wedge (a_1 v^1 \wedge \cdots \wedge a_{k-1} v^{k-1}).$$

Now the length of

$$a_1 v^1 \wedge \cdots \wedge a_{k-1} v^{k-1}$$

is 1 over the Jacobian of the nearest point projection to C, taken at z. Since C is a cone, this is the same as taking the Jacobian of the nearest point projection to B at $\frac{1}{r \cos \theta} z$. This is worked out in [W, p. 464, line 10], and is

$$(\det(I - \|\frac{z}{x} - p\| \, \mathbf{h}_{ij}^\nu))^{-1}$$

$$= (\det(I - \tan(\theta) \, \mathbf{h}_{ij}^\nu))^{-1}$$

$$= (1 - O(\tan^2(\theta)))^{-1},$$

where $\mathbf{h}_{ij}$ is the second fundamental form of B at p, and $\mathbf{h}_{ij}^\nu$ is the $k-1$ by $k-1$ matrix obtained by projecting $\mathbf{h}_{ij}$ onto the unit normal vector ν (now with base point p). (Note: In Weyl's notation, let our ν equal $n(1)$, and let $t_1 = 1$ and $t_i = 0$ for $i > 1$; then $G_{\alpha\beta}(1) = - < \nabla_{f_\alpha}(f_\beta), \nu >$, $G_{\alpha\beta}(i) = 0$ for $i > 1$, and $\|g_{\alpha\beta}\| = 1$).

Therefore we now have

$$\frac{\text{Vol}(E)}{\text{Vol}(F)} = \frac{\sqrt{(dr/d\theta)^2 + r^2}}{r^{2k+2}\cos^{2k-2}\theta \, \det(I - \tan(\theta)\,\mathbf{h}_{ij}^{\nu})}.$$

This will mean that the projection curves must have less curvature, so the normal wedge will be wider. In particular, the curve must now satisfy

$$\frac{dr}{d\theta} \leq r\sqrt{r^{2k}(\cos\theta)^{2k-2}\inf_{\nu}(\det(I - \tan(\theta)\,\mathbf{h}_{ij}^{\nu}))^2 - 1}$$

$$r(0) = 1.$$

We have replaced the determinant by its infimum over all normal directions ν, because we want to use the same curve for all directions radiating from p.

Now if the curvature is too large, then the argument under the radical will be negative before $r \to \infty$, and we will fail to get the desired retraction.

We summarize the above derivation in the following definition:

1.1.6 Definition: Let $p \in B \subset \mathbf{S}^{n-1}$, and $\dim(B) = k - 1 \geq 2$. Let $\mathbf{h}_{ij}^{\nu}$ be the matrix representing the second fundamental form of B at p, projected in the normal direction ν. Define the projection curve γ_p, if it exists, as the "narrowest" curve satisfying the differential equation

$$\frac{dr}{d\theta} = r\sqrt{r^{2k}(\cos\theta)^{2k-2}\inf_{\nu}(\det(I - \tan(\theta)\,\mathbf{h}_{ij}^{\nu}))^2 - 1}$$

$$r(0) = 1.$$

Existence depends on the curvature $\mathbf{h}$ being small enough, and\or the dimension being large enough. If the curvature is small, there is a one-parameter family of solution curves (see Chapter 3); by "narrowest" we mean the one in which r grows fastest as a function of θ.

If γ_p exists, it will either go to infinity as θ approaches some finite value, or $dr/d\theta$ will vanish at some positive θ. In the latter case, we fail to prove that C is minimizing; in fact, for some types of cones we can then deduce that C is stable but not area-minimizing (see chapter 4). In the case where $r \to \infty$ as $\theta \to \theta_0$, we call $\theta_0 = \theta_0(p)$ the "vanishing angle" at p, because $\theta_0(p)$ defines the width of the normal wedges, outside of which the retraction Π sends everything to $\mathbf{0}$.

1.1.7 Definition: For $p \in B \subset \mathbf{S}^{n-1}$, define the "vanishing angle" $\theta_0(p)$, if it exists, as the limit of θ as $r \to \infty$ in the curve γ_p defined above.

Note: An equivalent definition of the vanishing angle is given in Chapter 2, the chapter on calibrations (see 2.3.6). The second definition, also in terms of an O.D.E., has better numerical properties for solving by computer.

1.1.8 Terminology: If $C = 0 \ast B$ is a cone for which the vanishing angles $\theta_0(p)$ exist for each $p \in B$, we will often refer to "the normal wedges of C," meaning the normal wedges $W_p(\theta_0(p))$, which have the particular radii $\theta_0(p)$.

Section 1.2: The curvature criterion

We are now ready to give the first version of the main theorem of this paper. It is slightly stronger than the second version, Theorem 1.3.5, but more difficult to apply directly.

1.2.1 Theorem (A "Curvature Criterion"): Let B be a smooth minimal $k-1$ dimensional submanifold of $\mathbf{S}^{n-1}$, not necessarily orientable, and $C = 0 \times B$, the cone over B. Let $\theta(p)$ be the vanishing angle function of B, as defined above, if it exists. For each $p \in B$, let $W_p(\theta(p))$ be the normal wedge described in Definition 1.1.2. If $\theta(p)$ exists for all p and no two of these normal wedges intersect, then C minimizes area in the class of oriented and unoriented surfaces.

Further, if a solution of 1.1.6 exists at least on a small interval $[0, \theta]$, then C is stable, in the sense that a small perturbation of $\mathbf{R}^n$ which holds B and a neighborhood of 0 fixed cannot decrease the area of the truncated cone C_1.

Proof: Define a retraction $\Pi : \mathbf{R}^n \to C$ as in Definition 1.1.5, using the curves γ_p prescribed in Definition 1.1.6. Since the normal wedges do not intersect, Π is well-defined. By the defining property of γ_p, Π does not increase the area (k-volume) of any k-dimensional surface element.

Suppose first that B is orientable. Let C_1 be the truncated cone $C \cap \mathbf{B}^n(1)$, whose boundary is B. Suppose that S is some other surface (integral current) with boundary B. Since Π is continuous, $\Pi(S)$ is a compact surface contained in C and having boundary equal to B. By a variant of the constancy theorem [see M4, Theorem 4.9], if we let C_0 be the manifold given by deleting the origin and

B from C_1, $\Pi(S) - C_1$ must have constant multiplicity ℓ within C_0. But then $\ell B = \partial(\Pi(S) - C_1) = 0$, so that $\ell = 0$, and thus $\Pi(S) = C_1$.

Since Π does not increase area,

$$\text{Area}(S) \geq \text{Area}(\Pi(S)) = \text{Area}(C_1).$$

Now if B is not orientable, we need to work with currents modulo 2. More generally, suppose we wish to show that C is minimizing modulo m, with $m \geq 2$. We require that B have multiplicity $\mu \leq m/2$.

Represent B as an integral current whose boundary is zero modulo m, and as before, let $C = 0 \divideontimes B$. Let S be a current modulo m with $\partial S = B$ mod m, and let T be an integral current which is a representative of S modulo m (see [F, 4.2.26]). We wish to show that

$$\Pi(T) = C \quad \text{mod } m.$$

Let

$$D = 0 \divideontimes \partial(\Pi(T) - C).$$

Then

$$\partial(\Pi(T) - C - D) = 0$$

so that again by the constancy theorem, $\Pi(T) - C - D$ has constant multiplicity which can only be zero;

$$\Pi(T) - C - D = 0$$

$$\Pi(T) - D = C.$$

The fact that

$$\partial(\Pi(T) - C) = 0 \quad \text{mod } m$$

implies that

$$D = 0 \quad \text{mod } m$$

so that

$$\Pi(T) = C \quad \text{mod } m.$$

Finally, if a solution of 1.1.6 exists on an interval $[0, \theta]$, then C is stable, for we can define Π on a small cone-shaped neighborhood of C. Any surface lying entirely in this neighborhood and having boundary B must have at least as much area as C_1. ∎

Section 1.3: Simplifications

Next we will simplify the hypotheses of Theorem 1.2.1. This will weaken the theorem somewhat, but will make it easier to apply.

Nonintersection of normal wedges

First, it is difficult to directly calculate whether or not any of the normal wedges intersect. We will give a stronger condition which is easier to use, and which is equivalent in the case of isoparametric cones whose normal radius (see Definition 1.1.3) is the same everywhere. The condition is in terms of the normal radius, which is easier to calculate.

1.3.1 Lemma: If α is less than or equal to the minimum focal distance of C, and the normal radius of C is at least 2α everywhere (i.e., no normal wedge of radius 2α intersects the cone except along the wedge's center ray), then no two normal wedges of radius α intersect each other.

Note: If the vanishing angle exists at a point p, it is automatically less than or equal to the focal distance of C at p. Otherwise, by definition of focal distance, normal wedges near p would intersect each other, making the nearest-point projection undefined at that radius; thus, the vanishing angle could not exist at p. This tells us, for example, that if C has the same curvature at each point $p \in B$, and the vanishing angle is α, then we need not calculate the focal distance to apply Lemma 1.3.1.

Proof of Lemma 1.3.1: The idea is to take the *shortest* pair of distinct (closed) normal geodesics which intersect at a point $x \notin M$, and show that they must be parallel at x, so that their union is a normal geodesic intersecting M in two places. (Here, "shortest pair" means that the longer of the two is as short as possible.)

First note that since the normal wedges are not closed at their boundaries, it is equivalent to prove Lemma 1.3.1 for α *less than* the minimum focal distance. This stronger hypothesis makes the proof a little easier. Let

$$\beta = \inf_{x \in \mathbf{S}^n} (\max\{\mathrm{dist}_1(x, M), \mathrm{dist}_2(x, M)\})$$

where dist_1 and dist_2 are the lengths of normal geodesics starting at *distinct* points p and q in M, and ending at x. Take a subsequence converging to the infimum, such that x_i, p_i, and q_i converge to x, p, and q; then

$$\max\{\text{dist}(x,p), \text{dist}(x,q)\} = \beta.$$

Suppose that $\beta \leq \alpha$, thus contradicting the conclusion of the lemma. Then $p \neq q$, since β is less than the focal distance. Let σ_p be the geodesic arc from p to x, and σ_q the geodesic from x to q. The claim is that σ_p and σ_q are parallel at x, so that we can join them to get a single normal geodesic of length 2α or less, which intersects M in two places, thus contradicting the hypothesis of the lemma.

To see that σ_p and σ_q are parallel at x, first note that their lengths are *both* equal to β. If not, then WLOG, $\text{length}(\sigma_p) < \text{length}(\sigma_q)$. Then for some $\beta' < \beta$, x is in the interior of the neighborhood $\mathcal{N}(M, \beta') \subset \mathbf{S}^n$. Inside this neighborhood, choose $x' \neq x$ on σ_q. Thus there would be distinct normal geodesics from M to x' of length less than β, which cannot happen.

By a similar argument, $x \notin \mathcal{N}(M, \beta)$, so $x \in \text{bd}\,(\mathcal{N}(M, \beta))$. Let Ω_p and Ω_q be small neighborhoods in M of p and q. Since β is less than the focal distance, $\text{bd}\,\mathcal{N}(\Omega_p, \beta)$ and $\text{bd}\,\mathcal{N}(\Omega_q, \beta)$ are smooth at x. By the definition of β,

$$\mathcal{N}(\Omega_p, \beta) \bigcap \mathcal{N}(\Omega_q, \beta) = \emptyset,$$

so $\text{bd}\,\mathcal{N}(\Omega_p, \beta)$ and $\text{bd}\,\mathcal{N}(\Omega_q, \beta)$ are tangent at x. Since the geodesics σ_p and σ_q are perpendicular to these $n - 1$ dimensional boundaries, they are parallel to each other. ∎

Lower bounds for $\det(I - \mathbf{h}^{\nu}_{ij})$

A second simplification involves the quantity $\inf_{\nu}(\det(I - t\,\mathbf{h}^{\nu}_{ij}))$. The infimum will be easier to estimate if we bound the determinant in terms of the Euclidean norm of the matrix. If C is codimension 1, this simplification is often not used, since there are only two normal directions from each point of C.

1.3.2 Lemma: Let $m \geq 2$. Let A be a $m \times m$ symmetric matrix whose trace is zero, and let $\alpha = \|A\| = (\sum a_{ij}^2)^{1/2}$. Define

$$F(\alpha, t, m) = \left(1 - \alpha t \sqrt{\frac{m-1}{m}}\right)\left(1 + \frac{\alpha t}{\sqrt{m(m-1)}}\right)^{m-1};$$

note that

$$F(\alpha, t, m) = 1 - \frac{\alpha^2}{2}t^2 + O(t^3).$$

Then for all $t \in [0, \frac{1}{\alpha}\sqrt{\frac{m}{m-1}}]$,

$$\det(I - tA) \geq F(\alpha, t, m).$$

This inequality is sharp; for certain matrices, equality holds for all t.

Proof: We can assume A is diagonal. Otherwise, we can choose an orthogonal matrix Q such that $QAQ^T = \Lambda$ is diagonal, and

$$\det(I - tA) = \det(Q(I - tA)Q^T) = \det(I - t\Lambda),$$

while $\|\Lambda\| = \|A\|$, and $\mathrm{tr}\,\Lambda = \mathrm{tr}\,A$.

Thus, we need to minimize $\prod(1 - ta_{ii})$ subject to $\sum a_{ii} = 0$ and $\sum a_{ii}^2 = \alpha^2$. This is done in Corollary A2, in the appendix. ∎

1.3.3 Corollary: For $m \geq 2$ and $t \in [0, \frac{1}{\alpha}\sqrt{\frac{m}{m-1}}]$,

$$\det(I - t\,\mathbf{h}_{ij}^\nu) \geq F(\alpha, t, m),$$

where $\alpha = \sup_\nu \|\mathbf{h}_{ij}^\nu\|$.

Proof: For $m \geq 2$ and $\alpha, t > 0$, the function $F(\alpha, t, m)$ is a decreasing function of α. To see this, let $s = t\sqrt{\frac{m-1}{m}}$; then

$$F(\alpha) = (1 - \alpha s)(1 + \frac{\alpha s}{m - 1})^{m-1}$$

$$\frac{\partial F}{\partial \alpha} = -s(1 + \frac{\alpha s}{m - 1})^{m-1} + s(1 - \alpha s)(1 + \frac{\alpha s}{m - 1})^{m-2}$$

$$= s(1 + \frac{\alpha s}{m - 1})^{m-2}((1 - \alpha s) - (1 + \frac{\alpha s}{m - 1})) < 0.$$

Thus, we use $\alpha = sup_\nu \|\mathbf{h}_{ij}^\nu\|$ in Lemma 1.3.2 in order to bound $\inf_\nu(\det(I - t\,\mathbf{h}_{ij}^\nu))$. ∎

1.3.4 Corollary: Let m and α be as in Corollary 1.3.3 above, and $t \in [0, \frac{1}{\alpha}]$. Then

$$\det(I - t\,\mathbf{h}_{ij}^\nu) \geq (1 - \alpha t)e^{\alpha t}.$$

Proof: By Corollary A3 in the appendix, the function $F(\alpha, t, m)$ is a nonincreasing function of m. Its limit as $m \to \infty$ is $(1 - \alpha t)e^{\alpha t}$.   ∎

The above corollary is important because it is independent of m.

To further facilitate the application of our main theorem, we can now build a table of vanishing angles in terms of the sup norm of the 2nd fundamental form $\mathbf{h}_{ij}^{\nu}$. Using this table, which is found on pages 20-21, we now state a simplified (weaker) version of Theorem 1.2.1:

Simplified version of the minimization test

1.3.5 Theorem: Let B be a smooth minimal $k - 1$ dimensional submanifold of $\mathbf{S}^{n-1}$, not necessarily orientable, and $C = 0 \ast B$, the cone over B. Let N be the maximum value taken by the Euclidean norm of the second fundamental form

$$N = \max_{q} \left(\sup_{\nu} \left(\sum_{i,j}(\mathbf{h}_{ij}^{\nu})^2\right)^{1/2}\right),$$

as ν ranges over all unit normals to C at a point q, and q ranges over B. Use Table 1.4.1 (and Proposition 1.4.2 if necessary), with $\alpha^2 = N^2$ and $\dim(C) = k$, to find the vanishing angle θ.

Now let M be the minimum value of the normal radius of C, as described in Definition 1.1.3. If θ exists and $2\theta \leq M$, then C is area-minimizing in the class of orientable and unorientable surfaces.

Proof: This is a corollary of Theorem 1.2.1, using Corollaries 1.3.3 and 1.3.4 to calculate Table 1.4.1, and using Lemma 1.3.1 to show that the normal wedges do not intersect.   ∎

Section 1.4: Table of vanishing angles

The table on the following pages gives the vanishing angle, in degrees, of a cone C at a point $p \in C$, as a function of the curvature and dimension of C. The following notes will be helpful in using the table:

(1) The table was made using the bound in Corollary 1.3.3 for dimensions 3 to 11. To make possible the application of Proposition 1.4.2, the slightly weaker bound

in Corollary 1.3.4 was used in dimension 12. The numerical analysis is described in Section 3.4.

(2) The asterisks *** indicate that a solution of the O.D.E. exists in some interval $[0, \theta]$, but r does not go to infinity, so the vanishing angle does not exist. This information is useful for proving that certain cones are stable but not minimizing (see Theorems 4.3.1, 4.4.1, and 4.4.5).

(3) For any given cone, the bound in Corollary 1.3.3 may or may not be attained; the determinant may differ from F in the terms of order 3 or higher. This means that the vanishing angle for a cone may be smaller than that given in the table. By the same token, if the table lists "***" (see note 2), the vanishing angle *may* still exist for the cone in question. In such a case, to get further information one must solve the differential equation 1.1.6 (or 2.3.6) numerically from scratch, rather than using the table; see Section 3.4.

(4) The value to use on the left column is

$$\alpha^2 = (\sup_\nu \|\mathbf{h}_{ij}^\nu\|)^2.$$

The table has been made using the *square* of the sup norm, because this will often be a rational number.

(5) If $\dim(C) > 12$, use Proposition 1.4.2. Direct numerical calculation of the vanishing angle, using Definition 1.1.6, would yield a slightly smaller angle, but in most cases this would not be necessary.

(6) The values in the table are rounded **up** to the next hundredth of a degree.

(7) The values in parentheses are tentative, pending more precise numerical analysis. It appears that in dimensions 3 and 4, a cone satisfying the hypotheses of Corollary 4.4.7 (for which the curvature criterion is both necessary and sufficient) would either be unstable or area-minimizing. In particular, for $k = 3$ or $k = 4$, if there exists such a cone which is "marginally stable," i.e., $\alpha^2 = (\frac{k-2}{2})^2$, then the cone (apparently) must be area-minimizing.

Dimension of C

α^2	3	4	5	6	7	8
0.00	39.81	27.46	20.10	17.02	14.32	12.34
0.05	40.60	27.61	21.07	17.05	14.32	12.36
0.10	41.56	27.78	21.14	17.08	14.35	12.37
0.15	42.80	27.97	21.19	17.12	14.37	12.38
0.20	44.57	28.16	21.26	17.14	14.39	12.40
0.25	(50)	28.36	21.33	17.18	14.41	12.40
0.30	---	28.58	21.40	17.20	14.42	12.42
0.35	---	28.83	21.48	17.24	14.43	12.43
0.40	---	29.07	21.55	17.28	14.45	12.44
0.45	---	29.33	21.62	17.31	14.47	12.45
0.50	---	29.63	21.70	17.34	14.49	12.45
0.55	---	29.94	21.78	17.38	14.51	12.47
0.60	---	30.29	21.88	17.42	14.53	12.48
0.65	---	30.66	21.97	17.45	14.54	12.49
0.70	---	31.09	22.06	17.49	14.56	12.51
0.75	---	31.57	22.15	17.53	14.59	12.51
0.80	---	32.15	22.24	17.57	14.60	12.53
0.85	---	32.85	22.35	17.61	14.62	12.53
0.90	---	33.74	22.44	17.65	14.65	12.55
0.95	---	35.05	22.55	17.69	14.68	12.57
1.00	---	(41)	22.66	17.74	14.69	12.57
1.10	---	---	22.89	17.81	14.74	12.60
1.20	---	---	23.15	17.90	14.78	12.62
1.30	---	---	23.43	17.10	14.82	12.65
1.40	---	---	23.73	18.09	14.87	12.68
1.50	---	---	24.08	18.20	14.91	12.71
1.60	---	---	24.47	18.31	14.96	12.73
1.70	---	---	24.93	18.41	15.01	12.75
1.80	---	---	25.45	18.52	15.07	12.78
1.90	---	---	26.10	18.66	15.12	12.82
2.00	---	---	26.97	18.77	15.17	12.84
2.20	---	---	30.73	19.06	15.29	12.90
2.40	---	---	---	19.36	15.40	12.96
2.60	---	---	---	19.71	15.52	13.03
2.80	---	---	---	20.13	15.66	13.10
3.00	---	---	---	20.61	15.81	13.17
3.25	---	---	---	21.40	15.99	13.25
3.50	---	---	---	22.56	16.20	13.36
3.75	---	---	---	24.84	16.44	13.45
4.00	---	---	---	(***)	16.71	13.55
4.50	---	---	---	---	17.37	13.79
5.00	---	---	---	---	18.30	14.07
5.50	---	---	---	---	19.99	14.37
6.00	---	---	---	---	***	14.75
6.50	---	---	---	---	---	15.23
7.00	---	---	---	---	---	15.84
7.50	---	---	---	---	---	16.77
8.00	---	---	---	---	---	18.73
8.50	---	---	---	---	---	***
9.00	---	---	---	---	---	***

Dimension of C

a^2	9	10	11	12
0.00	10.86	9.69	8.75	7.97
0.20	10.88	9.70	8.76	7.99
0.40	10.91	9.73	8.78	7.99
0.60	10.94	9.74	8.79	8.01
0.80	10.97	9.77	8.81	8.01
1.00	11.01	9.79	8.82	8.03
1.20	11.04	9.81	8.83	8.04
1.40	11.07	9.83	8.85	8.05
1.60	11.10	9.85	8.87	8.06
1.80	11.14	9.87	8.88	8.07
2.00	11.17	9.90	8.89	8.08
2.50	11.26	9.96	8.94	8.12
3.00	11.36	10.03	8.98	8.15
3.50	11.46	10.09	9.03	8.18
4.00	11.57	10.16	9.07	8.21
4.50	11.71	10.23	9.13	8.25
5.00	11.83	10.32	9.17	8.28
6.00	12.14	10.48	9.29	8.36
7.00	12.51	10.68	9.41	8.45
8.00	12.99	10.91	9.54	8.55
9.00	13.68	11.18	9.69	8.64
10.00	14.83	11.51	9.86	8.75
11.00	***	11.94	10.05	8.87
13.00	---	13.51	10.55	9.17
15.00	---	***	11.35	9.55
17.00	---	---	13.90	10.11
19.00	---	---	***	11.23
21.00	---	---	---	***
23.00	---	---	---	***
25.00	---	---	---	***

For higher dimensions, the following principle is useful:

Suppose the vanishing angle has been estimated using the inequality in Corollary 1.3.4, as is the case for dimension 12 in Table 1.4.1. If we then double the dimension, we can also allow the sup norm α to be twice as big, and the tangent of the resulting vanishing angle will be less than half of the tangent of the previous angle. The same holds true for any other multiplying factor greater than 1, namely:

1.4.2 Proposition: Let $V_c(k, \alpha)$ denote the estimated vanishing angle which is computed using the inequality

$$\det(I - t\,\mathbf{h}_{ij}^{\nu}) \geq (1 - \alpha t)e^{\alpha t},$$

where $k = \dim(C)$ and $\alpha = \sup \|\mathbf{h}_{ij}^{\nu}\|$. (This is the inequality used in the computation of Table 1.4.1, for $k = 12$). Then for any $m > k$,

$$\tan(V_c(m, \tfrac{m}{k}\alpha)) < \frac{k}{m}\tan(V_c(k, \alpha)).$$

Proof: We refer the reader forward to Theorem 2.3.6, in which we give an equivalent form for the O.D.E. defining the projection curve γ_p. The differential equation is

$$\left(g(t) - \frac{t}{k}g'(t)\right)^2 + \left(\frac{g'(t)}{k}\right)^2 \leq \inf_{\nu}(\det(I - t\,\mathbf{h}_{ij}^{\nu}))^2$$

$$g(0) = 1.$$

The curve γ_p can then be defined in terms of g and g' (see Section 2.4). If g reaches 0 at $t = t_0$, then γ_p goes to infinity at $\theta = \tan^{-1}(t_0)$. According to the above hypothesis, we have actually solved the O.D.E.

$$\left(g(t) - \frac{t}{k}g'(t)\right)^2 + \left(\frac{g'(t)}{k}\right)^2 \leq \left((1 - \alpha t)e^{\alpha t}\right)^2$$

$$g(0) = 1.$$

Now set $t = \frac{m}{k}s$, and let

$$h(s) = g(t) = g(\frac{m}{k}s)$$

$$h'(s) = \frac{m}{k}g'(\frac{m}{k}s)$$

$$\left(h(s) - \frac{\frac{m}{k}s\,g'(\frac{m}{k}s)}{k}\right)^2 + \left(\frac{g'(\frac{m}{k}s)}{k}\right)^2 \leq \left((1 - \alpha\frac{m}{k}s)e^{\alpha\frac{m}{k}s}\right)^2$$

$$\left(h(s) - \frac{s\,h'(s)}{k}\right)^2 + \left(\frac{h'(s)}{m}\right)^2 \le \left((1 - (\frac{m}{k}\alpha)s)e^{(\frac{m}{k}\alpha)s}\right)^2$$

while $h' \le 0 \le h$ implies that

$$\left(h(s) - \frac{s\,h'(s)}{m}\right)^2 \le \left(h(s) - \frac{s\,h'(s)}{k}\right)^2.$$

Thus, h can be used in place of g to define γ_p. ∎

Chapter 2: Calibrations

Perhaps the most common method currently used for proving that a surface is area-minimizing is the method of calibrations. A good reference is [HL]; also see [M1] for an exposition on calibrations. The method itself can be described quite briefly; suppose that ϕ is a closed differential k-form of comass 1. That is, its maximum when applied to any oriented k-plane (identified with a unit simple k-vector) is 1. Suppose also that ϕ equals 1 when applied to any tangent plane of a particular surface M. Then M is area-minimizing. To prove this, suppose that $\partial N = \partial M$. Then by Stokes' theorem,

$$\text{Area}(M) = \int_M \phi = \int_N \phi \leq \text{Area}(N),$$

since $\phi = 1$ on M and $\phi \leq 1$ everywhere.

In this chapter, we use calibrations to derive the same results as we obtained in Chapter 1 (with the exception that we require a cone and its comparison surfaces to be orientable). There are three main reasons for including this chapter. First, the method of this paper was first discovered in the context of calibrations, rather than retractions. Thus, the motivation may be more clear. Second, I find it easier to derive the perturbation results in Chapter 6 using calibrations. Finally, it may be that future improvements of the method may come by doing a local construction of a *nonsimple* calibration on a cone (i.e., a form which takes its maximum on more than one k-plane at each point).

Section 2.1: Preliminary definitions

For a discussion of wedge products, k-vectors, and differential forms, see [M5, Section 4.1]. We define here a few terms.

2.1.1 Definitions: A *simple* differential k-form ϕ is one which, at a given point $z \in \mathbf{R}^n$, takes its maximum on a single k-plane ξ_z. The dual of ϕ at z is the k-plane ξ_z. The "pointwise comass" $\|\phi\|_z$ is this maximum value $\phi(\xi_z)$. The comass $\|\phi\|$ is the maximum over $\mathbf{R}^n$:

$$\|\phi\| = \sup\{\|\phi\|_z : z \in \mathbf{R}^n\}.$$

24

In this paper we will sometimes use the term "comass" to refer to "pointwise co-mass," relying on context to make the usage clear.

2.1.2 Definition: If ϕ is a differential k-form having comass 1, and $d\phi = 0$ (i.e., ϕ is closed), we call ϕ a calibration. If M is a surface such that $\phi = 1$ on every tangent plane of M, then M is "calibrated by ϕ". As noted above, a calibrated surface is area-minimizing.

Section 2.2: Zero curvature

We begin as in Chapter 1, with k-planes. We construct a singular calibration, which is homothetically invariant and is identically zero on part of $\mathbf{R}^n$. We will call this a "vanishing calibration."

Motivation

As motivation, consider the question of which m-tuples of k-planes through the origin in $\mathbf{R}^n$ are area-minimizing. We would like a condition that depends only on the relative position of any two of the planes (i.e., the angles between them.) Suppose we can find a singular calibration of a k-plane P which vanishes outside an "angular neighborhood" $\mathcal{N}$ of radius θ. (A point (x, z) lies in $\mathcal{N}$ if the

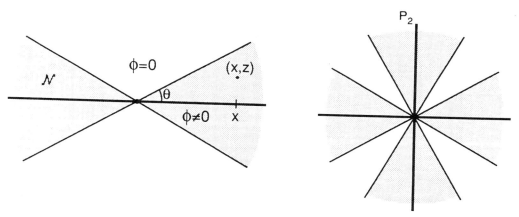

angular distance $\tan^{-1}(\frac{|z|}{r}) = \tan^{-1}(\frac{|z|}{|x|})$ is less than θ). Then if the angle between any pair of vectors lying in two different k-planes in our collection is at least 2θ, we can calibrate the whole union of k-planes by adding the vanishing calibrations corresponding to each plane. The sum of calibrations will still have comass 1, because wherever one of them is nonzero, the rest are zero.

Construction of a vanishing calibration

We will now see how to construct such a vanishing calibration on a k-plane, and afterwards apply the result to the problem of calibrating cones.

We start with the standard calibration

$$\omega = dx_1 \cdots dx_k.$$

Suppose we multiply ω by a function which is 1 on P and 0 outside the angular neighborhood $\mathcal{N}$. The resulting form will have comass 1, but will not be closed. Instead, we use the following method, which automatically gives a closed form:

(1) Since ω is closed, it is also exact. That is, there exist forms ψ such that $d\psi = \omega$. Choose a good representative. In particular, let

$$\psi = \frac{1}{k}(x_1\, dx_2 dx_3 \cdots dx_k - x_2\, dx_1 dx_3 \cdots dx_k + \cdots \pm x_k\, dx_1 \cdots dx_{k-1}).$$

Then $d\psi = \omega$, and $\omega = \frac{k}{r} dr \wedge \psi$.

(2) Multiply ψ by a function $g(t)$, which vanishes outside $\mathcal{N}$. We will take $t = \frac{|z|}{r}$ to be the tangent of the angular distance from P. Then we want $g(0) = 1$, $g'(0) = 0$, and $g(t) \equiv 0$ for $t \geq \tan\theta$, where θ is the angular radius of $\mathcal{N}$.

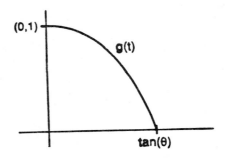

(3) Let $\phi = d(g\psi)$.

Notice that $g\psi$ is a Lipschitz form, but not C^1, because of the discontinuity at $\tan\theta$. Because $g\psi$ is Lipschitz, the discontinuous form $\phi = d(g\psi)$ can be used as a calibration; see Theorem A8 in the Appendix.

Now ϕ is automatically closed, because $d^2 = 0$. The condition that $g(0) = 1$ ensures that $\phi = 1$ on P. The requirement that ϕ have comass less than or equal to 1 at each point gives us a first order ordinary differential equation (or inequality) for $g(t)$ to satisfy. We compute

$$d(g\psi) = dg \wedge \psi + g\,d\psi$$

$$= \left(\frac{\partial g}{\partial z}dz + \frac{\partial g}{\partial r}dr\right) \wedge \psi + g\phi$$

$$= \left(\frac{g'}{r}dz - \frac{zg'}{r^2}dr\right) \wedge \psi + g\frac{k}{r}dr \wedge \psi$$

$$= \left((g - \frac{z}{rk}g')dr + \frac{g'}{k}dz\right) \wedge \left(\frac{k}{r}\psi\right)$$

$$= \left((g - \frac{t}{k}g')dr + \frac{g'}{k}dz\right) \wedge \left(\frac{k}{r}\psi\right).$$

Now dr, dz, and $\frac{k}{r}\psi$ have unit length and are orthogonal to each other. Thus, the equation $\|\phi\| = 1$ becomes

$$\left(g - \frac{t}{k}g'\right)^2 + \left(\frac{g'}{k}\right)^2 = 1.$$

Robert Bryant pointed out that this equation can be integrated (see Section 3.2 herein). The solutions which reach 0 exist only for $k \geq 3$. The (approximate) smallest value of θ for which $g(\tan\theta) = 0$ is given in the first line of Table 1.4.1. For example, for $k = 3$ it is approximately $39°$. As $k \to \infty$, the minimum θ approaches $\pi/(2k)$. See Section 5.3 for a further discussion of k-planes through a common vertex.

Section 2.3: Nonzero curvature

Now we turn to the question of calibrating more general cones. We construct the calibration locally, by putting local coordinates on a neighborhood of a point on the cone, and pulling back our vanishing calibration of a k-plane onto the cone.

Since the cone is not flat, it turns out that this increases the comass of the form; thus, we must start with a vanishing calibration of a k-plane whose comass decreases at a certain rate, as we move away from the plane.

We begin by defining a convenient way to extend coordinates.

2.3.1 Definition: Let M be a k-dimensional surface in $\mathbf{R}^n$. Let $e : U \subset \mathbf{R}^k \to M$ define coordinates on a regular neighborhood in M. We extend these coordinates to a neighborhood in $\mathbf{R}^n$, by letting $f = (e, e')$, where $e' = (e^{k+1}, \dots, e^n)$ is an isometry from $(x, \mathbf{R}^{n-k})$ to the normal space to M at $e(x)$.

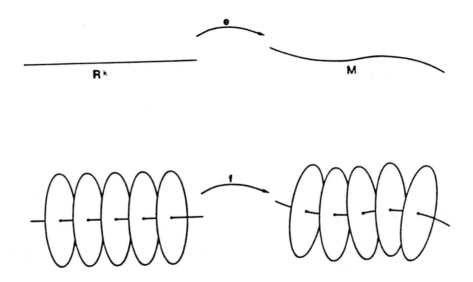

This involves a continuous choice of rotations of $\mathbf{R}^{n-k}$. If this is done in such a way that the coordinate map is $C^{1,1}$, then we call $f = (e^1, \dots, e^n)$ an "isonormal" extension of the coordinates $(e^1, \dots, e^k)$.

Note: The map f is a diffeomorphism when restricted to a small enough neighborhood so that its image does not include singularities or focal points of M; in practice, the set on which we are interested in using the isonormal extension will be sufficiently small.

Lemmas for computing the comass of a pull-back

We want to compute the differential of the nearest point projection π and of a one-sided inverse of π.

Let M be a k-dimensional surface in $\mathbf{R}^n$, and let p be a regular point on M. Put coordinates $\{e^i\}$ on M, centered at p, which are orthonormal at p. Let $(e^1, \ldots, e^n)$ be an isonormal extension of the coordinates to a neighborhood of p in $\mathbf{R}^n$. Let π be the nearest-point projection map onto M; thus,

$$\pi(e^1, \ldots, e^n) = (e^1, \ldots, e^k, 0, \ldots, 0).$$

Let σ_t^ν be a (one-sided) inverse of π, defined by

$$\sigma(p) = p + t\nu(p),$$

where $\nu(p)$ is a unit normal vectorfield to M.

2.3.2 Lemma: Let Y be a vector at $p + t\nu$ whose tangential component (parallel to T_pM) is T. The differential $D\pi$ is given by

(1) $$D\pi(Y) = \left[I - th_{ij}^\nu\right]^{-1}(T)$$

where h_{ij}^ν is the matrix whose (i,j) entry is $\langle \nabla_{e_i} e_j, \nu \rangle$ and T is written as a column vector in the basis $\{e_i\}$, and where the number t is smaller than the focal distance, that is, $\left[I - \tau h_{ij}^\nu\right]$ is nonsingular for all $\tau \in [0, t]$.

Now let $X = x^1 e_1 + \ldots + x^k e_k$ be a tangent vector to M at p. The differential $D\sigma$ is

(2) $$D\sigma(X) = X - t\left(\sum \langle \nabla_{e_i} e_j, \nu \rangle x^j e_i + N\right) = \left[I - th_{ij}^\nu\right]X - tN,$$

for some vector N normal to M at p, and perpendicular to ν.

Proof: Let γ_1 be a curve with initial velocity Y, and compute the tangent to the curve $\gamma_2 = \pi(\gamma_1)$. Thus, let

$$\gamma_1(s) = p + t\nu + sY,$$

and

$$\gamma_2(s) = \pi\big(\gamma_1(s)\big).$$

Let $X = \gamma_2'(0) = D\pi(Y)$. Since $\gamma_1(s) - \gamma_2(s)$ is a vector normal to M at $\gamma_2(s)$, we can write

$$\gamma_2(s) = \gamma_1(s) - tN(s),$$

with $N(0) = \nu(p)$. Differentiating by s, we get

$$X = \gamma_2'(0) = \gamma_1'(0) - tN'(0) = Y - t\nabla_x N$$

$$Y = X + t\nabla_x N$$

$$\mathrm{Tan}(Y) = X + t\mathrm{Tan}(\nabla_x N) = X + t\sum\langle \nabla_x N, e_j\rangle e_j.$$

Using the formula

$$X\langle Y, Z\rangle = \langle \nabla_x Y, Z\rangle + \langle Y, \nabla_x Z\rangle,$$

we get

$$\mathrm{Tan}(Y) = X - t\sum\langle \nabla_x e_j, N(0)\rangle e_j = X - \sum th_{ij}^{\nu} x^i e_j,$$

or in matrix notation,

$$\mathrm{Tan}(Y) = \Big[I - th_{ij}^{\nu}\Big]X.$$

Then

$$D\pi(Y) = X = \Big[I - th_{ij}^{\nu}\Big]^{-1}\mathrm{Tan}(Y).$$

For the second formula in the lemma, write

$$D\sigma(X) = T_1 + N_1.$$

Since

$$X = D\pi \circ D\sigma(X),$$

we have

$$X = \Big[I - th_{ij}^{\nu}\Big]^{-1}T_1,$$

or

$$T_1 = \Big[I - th_{ij}^{\nu}\Big]X.$$

We still need to know that N_1 is perpendicular to $\nu = \nu(0)$. For this, let $\gamma_2(s)$ be a curve with initial tangent vector X, and $\gamma_1(s) = \gamma_2(s) + t\nu(\gamma_2(s))$. Then

$$\gamma_1'(s) = \gamma_2'(s) + t\nabla_X \nu$$

$$\langle \nu, \gamma_1'(s)\rangle = t\langle \nu, \nabla_X \nu\rangle = 0,$$

since

$$0 = X\langle \nu(p), \nu(p)\rangle = \langle \nabla_X \nu, \nu\rangle + \langle \nu, \nabla_X \nu\rangle = 2\langle \nu, \nabla_X \nu\rangle. \quad \blacksquare$$

We will use this lemma to compute the comass of the pull-back of a differential form.

Suppose that ϕ is defined on a neighborhood $V \subset \mathbf{R}^n$. If we have a map $f : U \to V$, we can use f to map (pull back) ϕ to a form on U. We will need to know what is the dual of $f^\sharp(\phi)$, and what numerical value $f^\sharp(\phi)$ takes on this dual (this tells us the comass $\|f^\sharp(\phi)\|$). In general, if the dual of ϕ at p is ξ, the dual of $f^\sharp(\phi)$ at $f^{-1}(p)$ is *not* simply $Df^{-1}(\xi)$. Rather, we use the following method.

2.3.3 Lemma: Let $f : \mathbf{R}^n \to \mathbf{R}^n$ be a diffeomorphism. Let ϕ be a differential k-form which is simple at each point. Then $f^\sharp(\phi)$ is also simple, and we can compute its dual at a point by the following algorithm: Let ϕ be dual to the k-plane ξ_1 at the point p. Let

$$\xi_2 = v_1 \wedge v_2 \wedge \cdots \wedge v_{n-k}$$

be the orthogonal complement of ξ_1. Let $w_i = Df^{-1}v_i$. Then $f^\sharp(\phi)$ is dual to the orthogonal complement of $w_1 \wedge \cdots \wedge w_{n-k}$.

Proof: The conclusion is equivalent to saying that $f^\sharp(\phi) = 0$ on any k-plane containing any of the directions w_i. This is the same as saying that $\phi = 0$ on any k-plane containing any direction v_i, which is the same as ϕ being dual to ξ_1. $\quad \blacksquare$

Now we can use the previous two lemmas to compute the comass of a form which is constructed using isonormal coordinates.

2.3.4 Lemma: As in Lemma 2.3.2, let M be a k-dimensional surface in $\mathbf{R}^n$, let $e : U \subset \mathbf{R}^k \to M$, and let $f = (e^1, \dots, e^n)$ be an isonormal extension of e. Let π_1 and π_2 be the orthogonal projection maps onto $\mathbf{R}^k$ and M, respectively. Let $x \in \mathbf{R}^k$ and $p = f(x) \in M$. Suppose that the coordinates e^i are orthonormal at p.

Let $z \in \mathbf{R}^n$ be a point such that $\pi_1(z) = x$, and $\mathrm{dist}(z, x) = t$. Let $q = f(z)$; thus, $\pi_2(q) = p$.

Let ϕ be the form $e_1^* \wedge \cdots \wedge e_k^*$, defined on all of $\mathbf{R}^n$. Let ω be the pull-back $(f^{-1})^\sharp(\phi)$. Then the duals of ω at p and at q are parallel to the tangent plane $T_p(M)$, and

$$\left\|\omega\big|_p\right\| = 1$$

while

$$\left\|\omega\big|_q\right\| = \left(\det(I - th_{ij}^\nu)\right)^{-1}.$$

Proof: The orthogonal complement of the dual of ϕ is the normal space to $\mathbf{R}^k \subset \mathbf{R}^n$. The isonormal map f sends this to the normal space to M at p, and the orthogonal complement of this is $T_p(M)$.

To compute the comass,

$$\omega\big|_p(T_p(M)) = \phi\big|_x\left(Df^{-1}(T_p(M))\right) = 1,$$

since the coordinates $\{e^1, \ldots, e^k\}$ are orthonormal at p.

Now we compute $\|\omega\big|_q\|$. Let $v_1, \ldots, v_k$ be vectors at q, forming an orthonormal basis for the plane parallel to $T_p(M)$. Then we have

$$\omega\big|_q(T_p(M)) = \phi\big|_z\left(Df^{-1}(T_p(M))\right)$$

$$= \phi\left(Df^{-1}v_1 \wedge \cdots \wedge Df^{-1}v_k\right)$$
$$= \phi\left(D\pi_1 \circ Df^{-1}v_1 \wedge \cdots \wedge D\pi_1 \circ Df^{-1}v_k\right)$$

because ϕ does not measure components of $Df^{-1}v_i$ which are normal to $\mathbf{R}^k$;

$$= \left\|(D\pi_1 \circ Df^{-1})(v_1 \wedge \cdots \wedge v_k)\right\|$$
$$= \left\|(De^{-1} \circ D\pi_2)(v_1 \wedge \cdots \wedge v_k)\right\|$$
$$= \left\|D\pi_2(v_1 \wedge \cdots \wedge v_k)\right\|$$
$$= \left\|D\pi_2(v_1) \wedge \cdots \wedge D\pi_2(v_k)\right\|.$$

By Lemma 2.3.2, this is

$$\left\|Av_1 \wedge \cdots \wedge Av_k\right\| = \det A,$$

where A is the matrix

$$[I - th_{ij}^{\nu}]^{-1}. \quad \blacksquare$$

One consequence of this lemma is that $\omega|_p$ and $\omega|_q$ are independent of the isonormal extension of e.

Vanishing calibration of a cone

Now we are ready to calibrate C. Rather than constructing the calibration directly by pulling back a vanishing calibration of a k-plane, we will choose a different (but equivalent) method. Namely, our calibration of a k-plane was formed by taking $\phi = d(g\psi)$; instead of pulling back ω we will pull back ψ onto C, then describe a function g on $\mathbf{R}^n$ using the curvature of C, and take the exterior derivative of the (global) Lipschitz form $g \cdot (f^{-1})^{\sharp}(\psi)$.

Our first task is to pull back ψ by using local coordinates, and compute its comass. Recall that if $z \in \mathbf{R}^n$ and $x = \pi(z)$ is the nearest point to z in $\mathbf{R}^k \subset \mathbf{R}^n$, then ψ is the simple $k-1$ form whose duals at z and at x are the same (i.e., parallel), given by $T_x(\mathbf{S}^{k-1}(\|x\|))$ and having norm $\|x\|/k$.

2.3.5 Definition: Let $\{e^1, \ldots, e^{k-1}\}$ be coordinates mapping $U \subset \mathbf{S}^{k-1}$ to $V \subset B$; by a theorem of Moser [Ms] we can choose $\{e^i\}$ so that they preserve area at each point. Extend to coordinates on $0 \mathbin{\text{\ss}} V$ in the natural way, by letting $e^k = r$ and $e(ax) = ae(x)$, for $x \in U$. Let $f = \{e^1, \ldots, e^n\} : 0 \mathbin{\text{\ss}} U' \to 0 \mathbin{\text{\ss}} V'$ be an isonormal extension of $\{e^1, \ldots, e^k\}$, where U' is a neighborhood of U in $\mathbf{S}^{n-1}$, and V' is a corresponding neighborhood of V. For convenience, choose a homogeneous extension, that is, $f(az) = af(z)$.

Now let ψ be defined as in Section 2.2, and let

$$\tilde{\psi} = (f^{-1})^{\sharp}\psi.$$

2.3.6 Lemma: Let $q = f(z) \in \mathbf{R}^n$, and let $p \in C$ be the nearest point to q. Let $t = \text{dist}(p, q)$. The k-plane ξ dual to $\tilde{\psi}$ at p is contained in $T_p(C)$ and is perpendicular to the radial vector at p. The dual at q is parallel to the dual at p. Further,

$$\|\tilde{\psi}\|_p = \frac{|p|}{k}$$

while

$$\|\tilde{\psi}\|_q = \frac{|p|}{k}\det(I - t\,\mathbf{h}_{ij}^\nu),$$

where $\mathbf{h}_{ij}^\nu$ is the second fundamental form at p, projected onto the unit normal ν pointing from p toward q.

Proof: The direction of the duals is given by Lemma 2.3.3, using the fact that f preserves the normal space and the radial direction. The comass of $\tilde{\psi}$ at p is the same as the comass of ψ at $f^{-1}(p)$, since f preserves k-dimensional area on C. The comass of ψ at $z = f^{-1}(q)$ is also $|p|/k$; it remains to compute that the stretch factor in pulling back ψ is $\det(I - t\,\mathbf{h}_{ij}^\nu)^{-1}$ (which is the same as the Jacobian of the nearest-point projection to C at q).

Let $v_1,\ldots,v_{k-1}$ be vectors at q, forming an orthonormal basis for ξ. Let π_1 and π_2 be nearest-point projections onto $\mathbf{R}^k$ and C, respectively. Then we have

$$\tilde{\psi}\big|_q(\xi) = \psi\big|_z(Df^{-1}(\xi))$$

$$= \psi\big(Df^{-1}v_1 \wedge \cdots \wedge Df^{-1}v_{k-1}\big)$$

$$= \psi\big(D\pi_1 \circ Df^{-1}v_1 \wedge \cdots \wedge D\pi_1 \circ Df^{-1}v_{k-1}\big)$$

because ψ does not measure components of $Df^{-1}v_i$ which are normal to $\mathbf{R}^k$;

$$= \frac{|p|}{k}\big\|(D\pi_1 \circ Df^{-1})(v_1 \wedge \cdots \wedge v_{k-1})\big\|$$

$$= \frac{|p|}{k}\big\|(De^{-1} \circ D\pi_2)(v_1 \wedge \cdots \wedge v_{k-1})\big\|$$

$$= \frac{|p|}{k}\big\|D\pi_2(v_1 \wedge \cdots \wedge v_{k-1})\big\|$$

$$= \frac{|p|}{k}\big\|D\pi_2(v_1) \wedge \cdots \wedge D\pi_2(v_{k-1})\big\|.$$

In the next-to-last equality we used the fact that e is area-preserving. Now since this last expression does not depend on coordinates, we can assume that $\{e^i\}$ are actually orthonormal at p. Then by Lemma 2.3.2, we have

$$\tilde{\psi}\big|_q(\xi) = \frac{|p|}{k}\big\|Av_1 \wedge \cdots \wedge Av_{k-1}\big\| = \frac{|p|}{k}\det A,$$

where A is the matrix

$$\left[I - t h_{ij}^{\nu} \right]^{-1}. \quad \blacksquare$$

2.3.7 Corollary: The pull-back $\tilde{\psi}$ is independent of the choice of area-preserving coordinate map e and homogeneous isonormal extension f. Thus, we can define $\tilde{\psi}$ on a cone-shaped neighborhood of (all of) C by defining it locally as above. $\quad \blacksquare$

We will now state the main theorem of this chapter, which is equivalent to Theorem 1.2.5 (except for unorientable cones). In the proof we will use $\tilde{\psi}$ to define a calibration of C.

2.3.8 Theorem: Let B be an orientable C^2 submanifold of $\mathbf{S}^{n-1}$, and $C = 0 \times B$. Suppose that for each $p \in B$, there exists a function $g_p(t)$ satisfying $g_p(0) = 1$ and

$$(g_p - \frac{t}{k} g_p')^2 + (\frac{g_p'}{k})^2 \le \det(I - t h_{ij}^{\nu})^2$$

for all unit vectors ν normal to C at p. If $\theta(p)$ is a function on B such that $g_p(\tan \theta(p)) = 0$, and if the normal wedges of radius $\theta(p)$ do not intersect, then C is area-minimizing. $\quad \blacksquare$

Proof: Use the functions g_p to construct a function g defined on all of $\mathbf{R}^n$, as follows. If $z \in \mathbf{R}^n$ is in the normal wedge containing $q \in B$ and having radius $\theta(q)$, let t be the tangent of the angle between the vectors $\vec{0z}$ and $\vec{0q}$. Let $g(z) = g_q(t)$. If z lies outside the union of normal wedges of radius $\theta(p)$, then let $g(z) = 0$. Define

$$\phi = d(g \tilde{\psi}).$$

Since $g \tilde{\psi}$ is Lipschitz, ϕ is an acceptable candidate for a calibration (see Theorem A8, in the Appendix).

We will compute the comass of ϕ only at points $z \in \mathbf{R}^n$ such that the nearest point p in the cone lies in B. This will suffice because $\phi|_z = \phi|_{az}$, for $a > 0$, as will be apparent from the calculations to follow.

We begin by computing

$$d\tilde{\psi} = d(f^{-1})^{\sharp}\psi = (f^{-1})^{\sharp}d\psi = (f^{-1})^{\sharp}\omega$$

which, by computations similar to those we did for $(f^{-1})^{\sharp}\psi$, is equal to

$$\det(I - t\, \mathbf{h}_{ij}^{\nu})^{-1} \cdot T_p(C).$$

Then just as in calibrating a k-plane, we calculate

$$d\tilde{\psi} = \frac{k}{r} dr \wedge \tilde{\psi},$$

where dr is dual to the vector $\overrightarrow{0p}$ (rather than $\overrightarrow{0z}$).

Now we compute

$$d(g\tilde{\psi}) = dg \wedge \tilde{\psi} + g\, d\tilde{\psi}$$

$$= (\frac{\partial g}{\partial r} dr + \frac{\partial g}{\partial z} dz + v^*) \wedge \tilde{\psi} + \frac{k}{r} dr \wedge \tilde{\psi}$$

where dz denotes the dual of the unit vector pointing from p toward z, and v^* is dual to some vector lying in the $k - 1$ plane dual to $\tilde{\psi}$. Thus, the wedge product annihilates v^*. (This is why we need not assume that C is isoparametric).

By calculations similar to those of Section 2.2, we have

$$\phi = ((g - \frac{t}{k} g') dr + \frac{g'}{k} dz) \wedge \frac{k}{r} \tilde{\psi},$$

and

$$\|\phi\|^2 = ((g - \frac{t}{k} g')^2 + (\frac{g'}{k})^2) \det^2(I - t\, \mathbf{h}_{ij}^{\nu})$$

which equals 1 on C, and which by hypothesis is at most 1 at each point of $\mathbf{R}^n$. ∎

Section 2.4: Relationships between Chapter 1 and Chapter 2

Equivalence of the differential equations

Finally, let us prove that the differential equation in Theorem 2.3.6 and that of Chapter 1 (1.1.6) give equivalent results.

Let $\gamma = (r, \theta) = (f(\theta), \theta)$ be the projection curve given by (1.1.6); thus,

$$f'(\theta) = f(\theta)\sqrt{f^{2k}(\theta)(\cos^{2k-2}(\theta)\inf_{\nu}\det(I - t\mathbf{h}_{ij}^{\nu})^2 - 1}$$

and $f(0) = 1$. Set

$$g(t) = g(\tan\theta) = (f\cos\theta)^{-k}.$$

Then if f goes to infinity, g goes to zero. Noting that

$$\frac{dg}{dt} = \cos^2\theta \frac{dg}{d\theta},$$

a straightforward (if lengthy) calculation verifies that

$$(g - \frac{t}{k}\frac{dg}{dt})^2 + (\frac{1}{k}\frac{dg}{dt})^2 = \inf_\nu \det(I - t\mathbf{h}_{ij}^\nu)^2 \equiv p^2(t).$$

We can also compute that

(3)
$$\frac{g'/k}{g - \frac{t}{k}g'} = \frac{tf - f'}{f + tf'}.$$

For any curve $\gamma = (f(\theta), \theta)$ in polar coordinates, the slope (in terms of rectangular coordinates) of γ is

$$\frac{f + \tan\theta f'}{f' - \tan\theta f},$$

which is the negative reciprocal of (3). From this we have the following proposition.

2.4.1 Proposition: The dual of the calibration ϕ of Chapter 2, at any point where $\phi \neq 0$, is the orthogonal complement of the tangent plane to the surface of retraction S_p defined in Chapter 1 (see 1.1.5).   ∎

A relationship between retractions and calibrations

We include at this point another useful lemma which relates a calibration to a retraction whose surfaces of retraction are orthogonal to the duals of the calibration.

2.4.2 Lemma: Let ϕ be a simple closed k-form in $\mathbf{R}^n$. Let $\Pi : \mathbf{R}^n \to C$ be a retraction onto a k-dimensional surface. For each $p \in C$, let $S_p = \Pi^{-1}(p)$ be the "surface of retraction" passing through p. Suppose that for each p, at each point of S_p the dual of ϕ is completely orthogonal to S_p. Let P be a small piece of k-dimensional surface in $\mathbf{R}^n$, and let $E = \Pi(P)$. Then

$$\int_P \phi = \int_E \phi.$$

Proof: If C is codimension 1, then each S_p is a curve; let D be the union of the curves S_p extending from points of E to points of P (a "bent dowel" whose ends are E and P). By Stokes' theorem,

$$\int_{\partial D} \tilde\omega = 0.$$

Since the projection curves are orthogonal to the dual k-planes of $\tilde\omega$, $\tilde\omega = 0$ on all of ∂D except on E and P. If we orient E and P in the "same" way, this says that

$$\int_P \tilde\omega = \int_E \tilde\omega.$$

Now if C is not codimension 1, make a continuous choice of smooth curves $s_p \subset S_p$ for all $p \in P$ (for example, s_p could be a geodesic in S_p).   ∎

Chapter 3: The Differential Equation

Section 3.1: Overview

In this chapter we will study some of the properties of the ordinary differential equation

$$(1) \qquad \left(g(t) - \frac{t}{k}g'(t)\right)^2 + \left(\frac{g'(t)}{k}\right)^2 = p^2(t),$$

and the corresponding initial value problem $(1')$ in which

$$g(0) = 1.$$

The function p(t) is a given polynomial of the form

$$p(t) = 1 + p_2 t^2 + \cdots + p_{k-1}t^{k-1},$$

with $p_2 \leq 0$, which comes from the curvature of a cone; k is the dimension of the cone.

For future reference we note one fact about $p(t)$: The roots of p are the eigenvalues of a symmetric matrix, so they are all real. This means that if p is positive on an interval $(0, t_0)$, then p' is negative on the whole interval.

We will first examine qualitative properties of this differential equation. We will use one of these properties, namely that $g''(0)$ exists, in order to do the numerical analysis at the end of this chapter. The rest of the qualitative analysis is not essential for the results of the other chapters of this paper.

Our main objective is to see whether there is a solution satisfying $g(0) = 1$ (and thus $g'(0) = 0$), and $g(t_0) = 0$ for some $t_0 > 0$. If there is such a solution, then we may be able to use it to prove that a given cone (whose dimension is k and whose curvature is represented by $p(t)$) is area-minimizing. If there is no such solution $g(t)$, then given certain further restrictions we can prove that the cone is not area-minimizing (see Chapter 4).

We will see that if $|p_2| \leq (k-2)^2/8$ then there is a one-parameter family of solutions of (1) which satisfy $g(0) = 1$. They all lie between a certain barrier curve $\beta(t)$ and a smallest solution g_0 of the initial value problem.

38

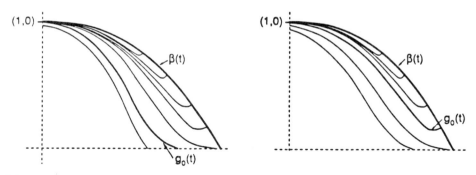

The solution g_0 may or may not reach 0. Any solution curve lying below g_0 must satisfy $g(0) < 1$. If $|p_2| > (k-2)^2/8$, then no solution of the O.D.E. satisfies $g(0) = 1$.

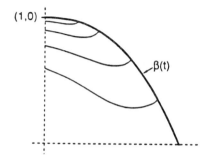

3.1.1 Note: There are actually two families of solutions to this differential equation, one of which we will ignore. If we use the quadratic equation to solve for $g'(t)$, we get the two solutions

(2)
$$g'(t) = \frac{k}{t^2+1}(tg + \sqrt{(t^2+1)p^2 - g^2})$$

and

(3)
$$g'(t) = \frac{k}{t^2+1}(tg - \sqrt{(t^2+1)p^2 - g^2}).$$

Thus, the barrier curve is defined by the requirement

$$(t^2+1)p^2 - g^2 \geq 0$$

$$g \le p(t)\sqrt{t^2 + 1}.$$

Through each point in the first quadrant lying below the barrier curve, there pass two solutions of (1). We are only concerned with the solution satisfying (3), because the solution of (2) will have positive derivative and cannot both exist for $t > 0$ and satisfy $g(0) = 1$. Throughout this chapter we will continue to use form (1) of the differential equation, with the implied restriction that we only allow solutions of (3).

Section 3.2: Zero curvature

First we will discuss the case when the curvature of the cone is zero, so that $p(t) \equiv 1$. Such a cone is a k-plane, or a union of k-planes through a common vertex. The analysis of this case is due to Robert Bryant.

In the equation

$$\left(g(t) - \frac{t}{k}g'\right)^2 + (\frac{g'}{k})^2 = 1,$$

let

$$g - \frac{t}{k}g' = \cos\theta,$$

$$\frac{g'}{k} = \sin\theta;$$

then

$$g - t\sin\theta = \cos\theta.$$

Differentiating by t, we get

$$g' - \sin\theta - t\cos\theta\frac{d\theta}{dt} = -\sin\theta\frac{d\theta}{dt}$$

$$k\sin\theta - \sin\theta - t\cos\theta\frac{d\theta}{dt} = -\sin\theta\frac{d\theta}{dt}$$

$$(k-1)\sin\theta = (t\cos\theta - \sin\theta)\frac{d\theta}{dt}.$$

Make the substitution

$$\sin\theta = -v^{k-1}.$$

Then

$$\cos\theta = \sqrt{1 - v^{2k-2}}.$$

Also,

$$\cos\theta \frac{d\theta}{dt} = -(k-1)v^{2k-2}\frac{dv}{dt}$$

so that

$$\frac{d\theta}{dt} = \frac{-(k-1)v^{k-2}}{\sqrt{1-v^{2k-2}}}\frac{dv}{dt}$$

$$-(k-1)v^{k-1} = (t\sqrt{1-v^{2k-2}} + v^{k-1})\frac{(1-k)v^{k-2}}{\sqrt{1-v^{2k-2}}}\frac{dv}{dt}$$

$$v = \left(t + \frac{v^{k-1}}{\sqrt{1-v^{2k-2}}}\right)\frac{dv}{dt}$$

$$\frac{v^{k-3}\,dv}{\sqrt{1-v^{2k-2}}} = \frac{v\,dt - t\,dv}{v^2} = d(\frac{t}{v}).$$

Thus

$$t(v) = v\left(c_0 - \int_0^v \frac{\rho^{k-3}\,d\rho}{\sqrt{1-\rho^{2k-2}}}\right)$$

$$g(v) = \cos\theta + t\sin\theta = \sqrt{1-v^{2k-2}} - v^k\left(c_0 - \int_0^v \frac{\rho^{k-3}\,d\rho}{\sqrt{1-\rho^{2k-2}}}\right).$$

If we take $c_0 = 0$, we get

$$t = \frac{-v^{k-1}}{k-2} + \cdots$$

$$g = 1 - \frac{v^{2k-2}}{2} - \frac{v^{2k-2}}{k-2} + \cdots = 1 - \frac{k}{2(k-2)}v^{2k-2} + \cdots$$

$$g(t) = 1 - \frac{k(k-2)}{2}t^2 + \cdots,$$

where the ellipsis in each case indicates higher-order terms.

If we take $c_0 \neq 0$, we find that the second-order coefficient of $g(t)$ is zero. Thus, all of the solutions $g(t)$ "cluster" around the solution $g \equiv 1$ except for one, $g_0(t)$, which has a negative second derivative at zero.

The solution $g \equiv 1$ corresponds to $c = \infty$.

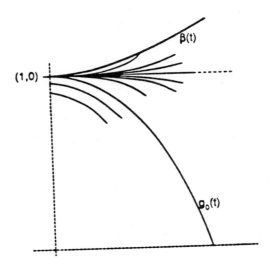

Note: In this case, the barrier curve $y = p(t)\sqrt{t^2 + 1} = \sqrt{t^2 + 1}$ does not come down to the t axis, as it does for nonzero curvature.

Section 3.3: General curvature

Now let us return to the more general differential equation

$$\left(g(t) - \frac{t}{k}g'\right)^2 + (\frac{g'}{k})^2 = (p(t))^2, \qquad g(0) = 1.$$

Again, we restrict our view to those solutions satisfying

(3) $$g'(t) = \frac{k}{t^2 + 1}(tg - \sqrt{(t^2 + 1)p^2 - g^2}).$$

We wish to prove that (if p has small second-order term) the picture of solutions is much like before; that there is a one-parameter family of solutions, that $g''(0)$ always exists and takes on two different values, and that exactly one solution g_0 has the larger curvature at zero.

Change of variables

Let

$$g - \frac{t}{k}g' = p(t)\cos\theta,$$

$$\frac{g'}{k} = p(t)\sin\theta;$$

then

$$g - tp(t)\sin\theta = p(t)\cos\theta.$$

Differentiating by t, we get

$$g' - p\sin\theta - tp'\sin\theta - tp\cos\theta\frac{d\theta}{dt} = p'\cos\theta - p\sin\theta\frac{d\theta}{dt}.$$

Gathering terms and dividing by $p\cos\theta$, we have

$$(k-1)\tan\theta - t\frac{p'}{p}\tan\theta + (\tan\theta - t)\frac{d\theta}{dt} = \frac{p'}{p}$$

or, letting $q(t) = \frac{p'(t)}{p(t)}$,

$$\frac{d\theta}{dt} = \frac{q(t)(1 + t\tan\theta) - (k-1)\tan\theta}{\tan\theta - t}.$$

Now we set

$$u = \tan\theta$$

and get

$$\frac{du}{dt} = \sec^2\theta\frac{d\theta}{dt} = (u^2 + 1)\frac{d\theta}{dt}$$

$$\frac{du}{dt} = (u^2 + 1)\frac{q(t)(1 + tu) - (k-1)u}{u - t}.$$

The initial condition is now

$$u(0) = \tan(0) = 0.$$

Though we don't yet know that $\frac{du}{dt}$ exists at 0, the above equation does hold for $t > 0$.

Let us assume for a moment that $u(t)$ is C^1 at $u(0) = 0$. Then

$$\lim_{t \to 0^+}\frac{u}{t} = \frac{du}{dt}\Big|_0.$$

In the above expression for $\frac{du}{dt}$, we divide top and bottom by t to get

$$\frac{du}{dt} = (u^2 + 1)\frac{(q/t)(1 + tu) - (k-1)(u/t)}{(u/t) - 1}.$$

Since

$$p(t) = 1 + p_2 t^2 + \cdots,$$

$$p'(t) = 2p_2 t + \cdots$$

$$q(t) = 2p_2 t + O(t^2)$$

$$\frac{q}{t} = 2p_2 + O(t).$$

Thus,

$$u'(0) = \frac{2p_2 - (k-1)u'(0)}{u'(0) - 1}.$$

Solving for $u'(0)$, we get

(4) $$u'(0) = \frac{2 - k \pm \sqrt{(k-2)^2 + 8p_2}}{2}.$$

This is all under the assumption that u is C^1 at $(0,0)$. It is apparent that u must not be C^1 at the origin if $(k-2)^2 + 8p_2 < 0$, for then $u'(0)$ would not be real. In fact, this is the only requirement. To see this, let us first turn the differential equation in u and t into an autonomous system of equations.

Autonomous system

Let $t = t(s)$ and $u = u(s)$, and set

$$\dot{u} = \frac{du}{ds} = (u^2 + 1)(q(t)(1 + tu) - (k-1)u)$$

$$\dot{t} = \frac{dt}{ds} = u - t.$$

The general theory of autonomous systems (see [BD], ch. 9, especially Thm. 9.2) tells us that we can understand this system near the origin by examining the behavior of the linearized system

$$\dot{u} = -(k-1)u + q_1 t$$

$$\dot{t} = u - t,$$

where $q_1 = 2p_2$, the first-order coefficient of the power series representation for $q(t)$. If $(k-2)^2 + 8p_2 > 0$, then the linearized system has integral curves passing through the origin, with slopes given by equation (4). In fact, there is a one-parameter family of integral curves through the origin. All but one have the same slope at

$(0,0)$, namely $\frac{1}{2}(2-k+\sqrt{(k-2)^2+8p_2})$. One "exceptional" integral curve passes through the origin with slope $\frac{1}{2}(2-k-\sqrt{(k-2)^2+8p_2})$. The theory tells us, then, that the nonlinear system has a family of solutions with these same characteristics. Further, these represent *all* solutions through the origin; they are all C^1 at 0.

3.3.1 Theorem: If $p(t)$ is such that $(k-2)^2+8p_2>0$, then all solutions of the initial value problem $(1')$ are C^2 at $(0,1)$.

Proof: By the theory of autonomous systems of differential equations, each solution $u(t)$ passing through the origin is C^1 there; since $u(t)=\tan(\theta(t))$ and $g'=kp(t)\sin(\theta(t))$, we deduce that $g'(t)$ is C^1 at 0, and thus $g(t)$ is C^2 at 0.

Now if $(k-2)^2+8p_2<0$, the theory tells us that solution curves wind around the origin in the (u,t) plane, rather than passing through the origin. This means that no C^1 solution of (1) remains in the first quadrant and satisfies $g(0)=1$; stated another way, we have the following result.

3.3.2 Theorem: If $(k-2)^2+8p_2<0$, then for any small $t_0>0$, there is an $\epsilon>0$ such that the solution curve of (1) satisfying $g(0)=1-\epsilon$ reaches the barrier curve $\beta(t)$ at $t=t_0$.

Proof: Let $y=(t_0,\beta(t_0))$. Starting at y, follow a solution γ of (1) back toward $t=0$. Since the corresponding curve in the (u,t) plane winds around the origin, $\gamma(0)<1$. ∎

Marginal case

Now suppose that $p(t)$ is such that $(k-2)^2+8p_2=0$. In this borderline case, the general theory does not tell us a priori whether any of the solutions $u(t)$ of the autonomous system pass through the origin; they may wind around 0. However, if we know that at least one integral curve passes through the origin, then none of the others could wind around 0; in this case the theory will again hold, and tell us that a one-parameter family of solutions pass through 0, all with the same slope $\frac{2-k}{2}$.

To see that one curve $u(t)$ passes through the origin, it is equivalent to prove that one solution $g(t)$ of the original O.D.E. satisfies $g(0)=1$. It is straightforward to show that if we choose a large enough positive number M and set

(5)
$$g(t)=1-\frac{k(k-2)}{4}t^2+Mt^3$$

then on some small interval $(0, t_0]$, g satisfies the inequality

(6)
$$\left(g(t) - \frac{t}{k}g'\right)^2 + (\frac{g'}{k})^2 < (p(t))^2.$$

Then by Lemma 3.4.1, there are also solutions of the initial value problem $(1')$.

3.3.3 Theorem: If $p(t)$ is such that $(k-2)^2 + 8p_2 = 0$, then the result stated in Theorem 3.3.1 still holds; the curves $g(t)$ passing through $(0,1)$ are C^2 at that point. ∎

An interesting consequence of (5) and (6) is the following. In Corollary 4.4.6 we will show that certain cones are unstable if $(k-2)^2 + 8p_2 < 0$. (Recall that $p_2 = \alpha^2/2$, where α is our sup norm defined in 1.3.3). This means that there are arbitrarily small perturbations of the cone which hold a neighborhood of the origin fixed, and decrease the area of the cone. Now if $(k-2)^2 + 8p_2 = 0$, then the cone is "marginally stable," in which case it is not clear whether some small perturbations might decrease its area. The following theorem states that they do not:

3.3.4 Theorem: Let C_1 be a (truncated) cone which satisfies the hypotheses of Theorem 4.3.1 or Theorem 4.4.1. (For example, C could be a minimal isoparametric codimension 1 cone.) As usual, let $p(t) = \det(I - t \, \mathbf{h}_{ij}^\nu)$. Suppose that $(k-2)^2 + 8p_2 = 0$, so that C_1 is "marginally stable." (Examples of such are not known). Such a cone would actually be "stable," in the sense that if a very small perturbation of the ambient space holds a neighborhood of the origin fixed, then the perturbed cone has **more** area than C_1.

Proof: We can use $g(t)$ as defined on line (5) above, to construct a calibration which exists in a small cone-shaped neighborhood of C, and away from C the calibration has comass less than 1. (C is the complete (nontruncated) cone). The standard calibration argument then tells us that surfaces lying inside the domain of definition of this calibration, with the same boundary as C_1, have more area than C_1.

Section 3.4: Numerical analysis of the differential equation

Since our results depend on numerical analysis, it is important to carefully examine our methods.

When we work with a differential equation numerically, we often need to examine some related differential inequalities. In this case, we wish to understand more fully the relationship between the equation

(1) $$\left(g(t) - \frac{t}{k}g'\right)^2 + (\frac{g'}{k})^2 = (p(t))^2$$

and the two inequalities

(2) $$\left(g(t) - \frac{t}{k}g'\right)^2 + (\frac{g'}{k})^2 \leq (p(t))^2,$$

and

(3) $$\left(g(t) - \frac{t}{k}g'\right)^2 + (\frac{g'}{k})^2 \geq (p(t))^2.$$

We will also use the modified equation (see Note 3.1.1) and inequalities

(1A) $$g'(t) = \frac{k}{t^2 + 1}(tg - \sqrt{(t^2 + 1)p^2 - g^2})$$

(2A) $$\frac{k}{t^2 + 1}(tg - \sqrt{(t^2 + 1)p^2 - g^2}) \leq g'(t) \leq \frac{k}{t^2 + 1}(tg + \sqrt{(t^2 + 1)p^2 - g^2})$$

(3A) $$g'(t) \leq \frac{k}{t^2 + 1}(tg - \sqrt{(t^2 + 1)p^2 - g^2}).$$

A prime after any of these, for example, $(1A')$, will indicate the initial value problem where we set $g(0) = 1$.

Note that $(2A)$ is equivalent to (2), while $(1A)$ and $(3A)$ are more restrictive than (1) and (3).

If a function $g(t)$ satisfies $(2')$ for $t \in [0, t_0]$, and if $g(t_0) \leq 0$, then we can use this g in Theorem 2.3.8, since a calibration is allowed to have comass less than 1 away from the area-minimizing surface. That is, for the purpose of proving that a cone is area-minimizing, we do not need an exact solution of (1), but only of (2). However, it is useful to know that this actually implies the existence of a solution of $(1')$.

3.4.1 Lemma: Suppose that the function $g_1(t)$ satisfies the inequality

$$\left(g(t) - \frac{t}{k}g'\right)^2 + (\frac{g'}{k})^2 < (p(t))^2$$

on a small interval $(0, t_0]$, and $g_1(0) = 1$. Then there exists a function $g_2(t)$ which satisfies

$$\left(g(t) - \frac{t}{k} g'\right)^2 + \left(\frac{g'}{k}\right)^2 = (p(t))^2$$

on an interval $[0, t_1]$, and $g_2(0) = 1$.

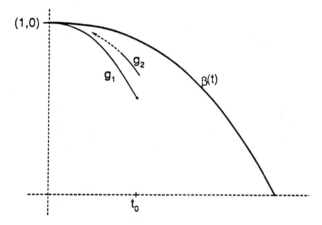

Proof: The function $g_1(t)$ must lie on or below the barrier $\beta(t)$. Since $\beta(t)$ itself does not satisfy the above inequality, $g_1 \not\equiv \beta$, so for some $t_1 < t_0$, $g_1(t_1) < \beta(t_1)$. Choose $g_2(t_1)$ such that $g_1(t_1) < g_2(t_1) < \beta(t_1)$, and define g_2 on $[0, t_1]$ by following an integral curve of (1A) backwards to $t = 0$. This curve cannot cross g_1, for at an intersection point, the equation (1A) and the inequality for g_1 would require that the slope of g_1 be greater than the slope of g_2, and thus they could not cross. Similarly, $g_2(t)$ cannot cross $\beta(t)$, so $g_2(0)$ must equal 1. ∎

On the other hand, suppose that we have a function g which satisfies (3A') and does not reach 0. Then no solution of the initial value problem (1') can reach 0.

3.4.2 Lemma: Let g_0 be the smallest solution of the 1-parameter family of solutions of the initial value problem (1A'). Write $g_0(t) = 1 - at^2 + O(t^3)$, with $a > 0$. Let g be another function with smaller second order coefficient, i.e., $g(t) = 1 - (a + \epsilon)t^2 + O(t^3)$. Suppose that for some interval $[0, t_0]$, g satisfies the inequality (3A') on $[0, t_0]$, and that $g'(t_0) = 0$ and $g(t_0) > 0$. Then g_0 does not reach 0.

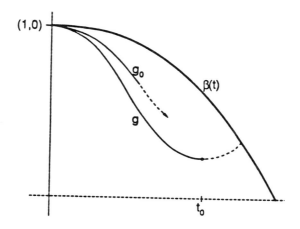

Proof: First note that since $g'(t_0) = 0$, $g(t_0) \geq p(t_0)$. Now extend g beyond t_0 by requiring that for $t > t_0$, g satisfies the *equation* (1A). Then g never reaches 0; in fact, g is increasing after t_0. To see this, note that

$$\frac{g'}{k} = \frac{gt - \sqrt{p^2(t)(t^2 + 1) - g^2}}{t^2 + 1} \geq 0$$

as long as $g \geq p$. Since p' is strictly negative between t_0 and the point where $\beta(t) = 0$, in order for g to be less than p for some t in this interval, g' would first have to be negative, which is a contradiction. On the other hand, g_0 and g can never cross, for at a crossing point we would have

$$g_0' < g' < \frac{k}{t^2 + 1}(gt - \sqrt{p^2(t)(t^2 + 1) - g^2},$$

a contradiction. Thus, g_0 never reaches 0.

Now that we have these lemmas, we can proceed with the numerical analysis. In order to use a numerical method for solving our O.D.E., we need to know how to "get started" near 0, since the equation is singular there. We want the smallest solution $g_0(t)$ passing through $(0, 1)$. Since we know that g_0 is C^2 at 0, we set

$$g_0(t) = 1 - at^2,$$

plug this into the O.D.E. (1), and solve for the larger value of a. This yields

$$a = \frac{k}{4}(k - 2 + \sqrt{(k - 2)^2 + 8p_2}).$$

We will use the function $g_0(t) = 1 - at^2$ for small t, to help us get a preliminary decision as to whether g_0 reaches 0. After this, we modify the method to get a more rigorous proof.

Preliminary method

To obtain the preliminary result, choose a small ϵ. A reasonable value may be $\epsilon = 10^{-3}$ or $\epsilon = 10^{-4}$. Let $g_0 = 1 - at^2$ for $t \in [0, \epsilon]$. Use the following implementation of Euler's method:

(1) Let $t = \epsilon$, $g = 1 - at^2$

(2) Let $g' = \frac{k}{t^2+1}\left(tg - \sqrt{(t^2 + 1)(p(t))^2 - g^2}\right)$

(3) If $g' \geq 0$ then quit; g failed to reach 0.

(4) Let $\delta = \epsilon t$

(5) Set $g := g + \delta g'$

(6) Set $t := t + \delta$

(7) If $g \leq 0$ then quit. Otherwise, loop back to step (2).

Notice that the value of δ is much smaller than t at each step. This is necessary.

The above method is quite reliable. It can be improved by setting $g = 1 - at^2 + a_3 t^3$, and solving for the correct value of a_3.

Examination of each linear segment of $g(t)$

For the purpose of a mathematical proof, (and the purpose of constructing Table 1.4.1), we must make the analysis more complete.

Our function $g(t)$ is piecewise linear. We specify g' on a linear segment by using the values of g and t at the left-hand endpoint of the segment. Thus, we need a bound on how fast the quantity

$$(g - \frac{t}{k}g')^2 + (\frac{g'}{k})^2 - p^2(t)$$

changes as we hold g' constant, increase t, and change g using the current value of g' to specify $\frac{\Delta g}{\Delta t}$. Take g' to be a constant, and differentiate by t:

$$2(g - \frac{t}{k}g')(g' - \frac{g'}{k}) - 2pp'$$

$$= 2(g - \frac{t}{k}g')(k - 1)(\frac{g'}{k}) - 2pp'.$$

Since $g - \frac{t}{k}g' \leq 1$, this derivative is bounded in absolute value by

$$2(k-1)|\frac{g'}{k}| + 2|pp'|.$$

If we take Δt to be very small compared to t, this will certainly be bounded on the whole interval from t_{old} to $t_{new} = t_{old} + \Delta t$, by the quantity

$$2|g'| + 3|p'(t_{old})|.$$

Getting started near $t = 0$

We also need a function to "get g started" near 0, which we can guarantee to satisfy the inequality (2) or (3A) for small t. We can get this by setting $g = 1 + at^2$ or $g = 1 + a_2 t^2 + a_3 t^3$, solving for a or for a_2 and a_3 so that g satisfies the O.D.E. (1) out to second or third order, and then either increasing or decreasing a or a_3 slightly. It may be a bit painful to prove that the inequality (2) or (3A) is actually satisfied on a particular small interval, so we provide here the following (also a bit painful) analysis:

We obtain g in a different way; we first find a function $u(t)$ as in Section 3, and then set $u = \tan\theta$ and $g = p\cos\theta + tp\sin\theta$.

We already calculated that if $u = -\alpha t + O(t^2)$, then to satisfy the differential equation, we want

$$\alpha = \frac{1}{2}(k - 2 + \sqrt{(k-2)^2 + 8p_2}).$$

Suppose now that we want $g(t)$ to satisfy the inequality (3A).

Choose a small tolerance number $E > 0$ ($E = 10^{-3}$ may be reasonable) and set

$$A = \alpha + E$$

$$u = -At$$

for small t. We want

$$\frac{du}{dt} < \frac{u^2 + 1}{u - t}(\frac{u}{t}(qt + 1 - k) + \frac{q}{t})$$

on our small interval. Thus, we want

$$-A < \frac{(A^2 t^2 + 1)}{-A - 1}(-A(qt + 1 - k) + \frac{q}{t}),$$

or

$$(*) \qquad 0 < A - \frac{(A^2 t^2 + 1)}{A+1}(-A(qt+1-k) + \frac{q}{t}).$$

Calculating this at $t = 0$, we get

$$C = A - \frac{A(k-1) + 2p_2}{A+1}.$$

If we estimate the derivative (by t) of the right hand side of $(*)$, we can see that for small t it is less than $3 + 10|p_3|$. It is more convenient to have this in terms of p_2; since p_2 and p_3 are coefficients of a polynomial with real roots, and $p_1 = 0$, it can be shown that $|p_3| \le |p_2|^{3/2}$. Thus, we can let $u = -At$ on the interval

$$[0, \epsilon] \equiv [0, \frac{C}{3 + 10|p_2|^{3/2}}].$$

Similarly, if we want $g(t)$ to satisfy the inequality (2), we set $A = \alpha - E$, calculate C as above, and let $u = -At$ on the interval

$$[0, \epsilon] \equiv [0, \frac{|C|}{3 + 10|p_3|}].$$

Now since $u = -At = \tan\theta$, we take

$$g = p\cos\theta + tp\sin\theta = p(t)\frac{1 - At^2}{\sqrt{1 + A^2 t^2}}.$$

Modified method

Once we have calculated A and ϵ as above, we use the following adaptation of our numerical method. The word "Adjust" denotes that we round *each* floating point calculation up or down slightly in order that computer round-off error not invalidate our result. To prove that g_0 does reach 0, "Adjust" rounds g up, H up (toward zero), $g + \delta g'$ up, and $t + \delta$ up. To prove that g_0 does not reach 0, we Adjust in the opposite direction (including rounding H *down* toward zero).

(1) Choose a small positive tolerance number E_1. A reasonable value may be 10^{-3} or 10^{-4}. Let $F = 1 + E_1$ (for proving that g_0 does not reach zero), or $F = 1 - E_1$ (for proving that it does).

(2) Let $t = \epsilon$ and $g = \text{Adjust}[p(t)\frac{1-At^2}{\sqrt{1+A^2t^2}}]$.

(3) Let

$$g' = F \cdot \left(\frac{k}{t^2+1}(tg - \sqrt{(t^2+1)(p^2(t)) - g^2}) \right)$$

(4) Calculate $H = \text{Adjust}[(g - \frac{t}{k}g')^2 + (\frac{g'}{k})^2 - p^2(t)]$. This will be slightly different from 0.

(5) If $g' \geq 0$ and $H \geq 0$, then quit.

(6) Let

$$\delta = \min\{ \frac{|H|}{2|g'| + 3|p'(t)|} , \frac{t}{1000} \}.$$

Then on the interval from t to $t + \delta$, the quantity calculated in line (5) will remain negative, or remain positive. Thus, either the inequality (2) or the inequality (3A) will be preserved on that interval.

(7) Set $g = \text{Adjust}[g + \delta g']$.

(8) Set $t = \text{Adjust}[t + \delta]$.

(9) If $g \leq 0$ then quit. Otherwise, loop back to step (3).

We can now use the lemmas we proved earlier in this section, which say that we can use differential inequalities to bound the solutions of the differential equation. If the above iteration confirms our initial decision as to whether or not the smallest solution g_0 of the initial value problem (1') reaches zero, we see that the initial information was, in fact, correct.

Chapter 4: Cones for Which the Criterion is Necessary as Well as Sufficient

Section 4.1: Introduction

Under certain conditions, the curvature criterion (Theorem 1.2.1) is necessary as well as sufficient. Thus, we can also prove that certain cones are *not* area-minimizing. Among the cones satisfying these extra conditions are the "isoparametric" codimension 1 cones, and cones over products of spheres.

Overview

The method of this chapter can be described as follows: Let $p \in B$. The fact that Theorem 1.2.1 fails will allow us (under certain circumstances) to define a certain vector field V on the normal wedge at p, such that there exists a (planar)

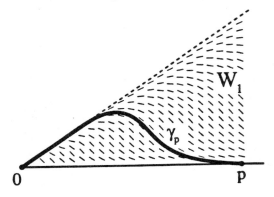

integral curve γ_p of V passing through p and the origin. The union of these curves γ_p, as p ranges over B, will form our comparison surface having less area than the truncated cone C_1.

Section 4.2: General setup

The first requirement we will make on B is that it have a certain symmetry.

4.2.1 Definition: Let B be a submanifold of the unit sphere, and let $C = 0 * B$. If the curvature (second fundamental form) is the same at each point of B, we call C an "isoparametric" cone.

54

More precisely, B is said to be isoparametric if, given any pair of points p and q in B, we can relate orthonormal bases of the tangent and normal spaces at p to corresponding orthonormal bases of the tangent and normal spaces at q, in such a way that the second fundamental form matrices $h_{ij}^{\nu}\big|_p$ and $h_{ij}^{\nu'}\big|_q$ are the same, for corresponding normals ν and ν'. In particular, a hypersurface H of the sphere is isoparametric if and only if the principal curvatures at one point of H (counting multiplicity) are the same as the principal curvatures at any other point.

In codimension 1, this is all we need to get a necessary and sufficient criterion for area minimization. In higher codimension, another thing we will need is the existence of a "conservative" normal vectorfield.

4.2.2 Definition: Let B be a submanifold of the unit sphere, and $C = 0 \divideontimes B$. Let $\nu(p)$ be a nonzero (and constant length) vectorfield defined on B, which is normal to C. If the derivative $\nabla_X \nu$ is tangent to B for each tangent vector X, then we will call $\nu(p)$ a "conservative" normal vectorfield.

If B is codimension 1 in $\mathbf{S}^{n-1}$, then any continuous unit normal vectorfield is automatically conservative, since there aren't any other normal directions in which ν could "wander." On the other hand, if B is of higher codimension then there may not even exist a *continuous* unit normal vectorfield. When we can choose a continuous unit normal vectorfield, we define the following method of constructing a surface with boundary B.

4.2.3 Definition: Suppose we can make a continuous choice $\nu(p)$ of unit normal to the cone $C = 0 \divideontimes B$, for all $p \in B$. Let $\gamma : [a, b] \to \mathbf{R}^2$ be a given fixed curve in the first quadrant, satisfying $\gamma(a) = (1,0)$ and $\gamma(b) = (0,0)$. Now for each $p \in B$, let γ_p be a copy of γ which we have embedded into $\mathbf{R}^n$ (into the normal wedge at p) by identifying the positive first coordinate axis of $\mathbf{R}^2$ with the ray $\overrightarrow{0p}$, and the positive second coordinate direction with the designated normal $\nu(p)$. We define the surface S induced by γ and $\nu(p)$ as the union of the curves γ_p, as p ranges over B. The orientation on S is induced by that of C, so that $\partial S = \partial C_1 = B$.

The following lemma gives us information about the tangent planes to these surfaces.

4.2.4 Lemma: Suppose that there exists a conservative unit normal vectorfield $\nu(p)$ on B. Let $\gamma(t)$ be a C^1 planar curve with endpoints $\mathbf{0}$ and p, as above, and let S be the surface induced by γ and $\nu(p)$. Let $p \in B$ and $t \in (a, b)$ be such that S is regular (and embedded) at the point $\gamma_p(t)$. We require that $\gamma_p(t)$ not be a focal point of C. Then the tangent plane to S at $\gamma_p(t)$ is $\gamma_p'(t) \wedge \xi$, where $\xi = T_p(B)$ is the $k - 1$ plane tangent to B at p.

Proof: S can be described as a union of points $q + \alpha(t)\nu(p)$, where p ranges over B, and each q is a point on the ray $\overrightarrow{0p}$, and α is a function depending only on the distance from 0 to q (which in turn depends on t). Clearly, if X is the radial tangent vector at q then $\nabla_X \gamma_p(t)$ is parallel to $\gamma_p'(t)$. If X is a tangent vector perpendicular to the radial vector, and thus parallel to the tangent plane $T_p(B)$, then

$$\nabla_X \gamma_p(t) = \nabla_X(q + \alpha(t)\nu(p)) = X + \alpha(t)\nabla_X \nu,$$

which is parallel to $T_p(B)$. Since $\gamma_p(t)$ is not a focal point of C, if $\{X^i\}$ is a basis for $T_p(B)$, then the vectors $X^i + \alpha(t)\nabla_{X^i} \nu$ are linearly independent, so they span $T_p(B)$. ∎

Section 4.3: Codimension 1

We are now ready to state the theorem for codimension 1 cones.

4.3.1 Theorem: Let B be a minimal, isoparametric $n - 2$ dimensional submanifold of $\mathbf{S}^{n-1}$, and $C = 0 \divideontimes B$. Suppose that Theorem 1.2.1 fails to prove that C is area-minimizing. Then in fact, C is not area-minimizing.

Proof: Let us see why Theorem 1.2.1 fails. A priori, there are two possibilities:

(1) A solution curve of the initial value problem in 2.3.6 exists and goes to zero, but the resulting normal wedges intersect.

(2) Either there is no solution of the initial value problem, or none of the solutions reach zero.

However, (1) cannot happen in this case. It is a general fact about minimal, codimension 1, isoparametric submanifolds of the sphere, that two normal wedges whose radii are smaller than the focal angle of C will never intersect. (A more familiar way of stating this is that the hypersurfaces of the sphere which are parallel to B are embedded (nonmimimal) isoparametric hypersurfaces, if the distance from B to the parallel hypersurface is less than the focal distance.) On the other hand, Theorem 1.2.1 never produces wedges that extend beyond the focal angle, because $p(t) = \det(I - t\, h_{ij}^{\nu}) = 0$ at that angle.

Thus, case (2) must have occurred.

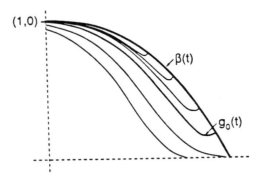

Suppose that there exist solution curves, but the smallest one hits the outer barrier $\beta(t)$ at a point (t_0, y_0) with $y_0 > 0$. Then we can choose a point (t_1, y_1) on $\beta(t)$ with $0 < y_1 < y_0$, for which a solution to the modified initial value problem

$$\left(g - \frac{t}{k}g'\right)^2 + \left(\frac{g'}{k}\right)^2 = (p(t))^2$$

$$g(0) = 1 - \epsilon$$

(for some $\epsilon > 0$) reaches the point (t_1, y_1). Call this curve g_1. Using g_1, we will construct a vector field in the normal wedge at p, and use an integral curve of this vector field to construct our comparison surface S.

Note 1: If no solution exists satisfying $g(0) = 1$, then we can start at *any* point (t_1, y_1) on $\beta(t)$ and follow a solution of the O.D.E. backwards to $t = 0$. Then $g(0)$ will be less than 1, so we can choose this curve as our solution g_1.

Note 2: There are two normals to the cone at each point; the polynomial $p(t)$ depends on which of the normals we let correspond to positive t. It can happen that solution curves $g(t)$ corresponding to one side of the cone do reach zero, while solution curves on the other side do not. If on *either* side of the cone C it happens

GARY R. LAWLOR

that there is no solution curve connecting $(0, 1)$ to the t axis, then C is not area-minimizing.

Now let W be a normal wedge, of radius $\tan^{-1}(t_1)$, and let W_1 be one half of W such that solutions g on that side of C do not reach 0.

For each point $z \in W_1$, let ϕ be the angle between the point (or vector) z and

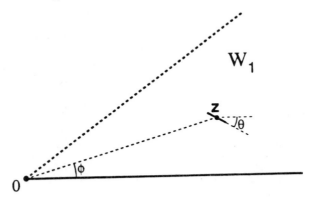

the horizontal. Let $t = \tan \phi$, and define a unit vectorfield V on W_1 which makes an angle θ with the horizontal, where θ is defined by

$$g_1 - \frac{t}{k} g_1' = p(t) \cos \theta,$$

$$\frac{g_1'}{k} = p(t) \sin \theta.$$

Since $g_1(0) < 1$, this vectorfield is not horizontal on the ray $\phi = 0$, which lies in the cone. Also, at the angle $\phi = \tan^{-1}(t_1)$, since (t_1, y_1) lies on $\beta(t)$ we have

$$\frac{g_1'(t_1)}{k} = \frac{t_1 g_1}{t_1^2 + 1}$$

so that

$$\tan \theta = \frac{\frac{g_1'}{k}}{g_1 - \frac{t_1}{k} g_1'}$$

$$= \frac{\frac{t_1 g_1}{t_1^2 + 1}}{g_1 - \frac{t_1^2 g_1}{t_1^2 + 1}} = t_1 = \tan \phi,$$

and thus $\theta = \phi$. Then an integral curve γ of V which starts at $p \in B$ remains within the closed wedge between the angles $\phi = 0$ and $\phi = \tan^{-1}(t_1)$, and goes to the origin.

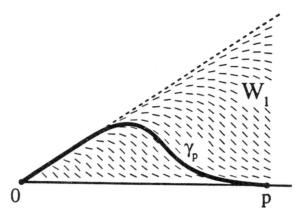

Now let S be the union of these curves γ for all $p \in B$, as described in Definition 4.2.3. Since $\tan^{-1}(t_1)$ is less than the focal angle, the closures of the normal wedges intersect only at 0, and thus S is embedded (and singular at 0).

Suppose now that B is orientable. Let ψ be the $k - 1$ form defined for C as in Section 2.2, and $\omega = d(g_1 \psi)$. Then ω is a calibration. By Lemma 4.2.4, the tangent planes to S are dual to ω, and thus S is calibrated. On the other hand,

$$\omega(C) = \big((d(g_1)) \wedge \psi\big)(C) + g_1\,(d\psi)(C) = g_1(0)\,(d\psi)|_0(C) = g_1(0) = 1 - \epsilon.$$

Thus,

$$\mathrm{Area}(S) = \int_S \omega = \int_{C_1} \omega = (1 - \epsilon)\mathrm{Area}(C_1).$$

Since ω is not defined on all of $\mathbf{R}^n$, we justify the use of Stokes' theorem by noting that the domain of ω contains the cone $D = 0 \divideontimes (S - C_1)$, whose boundary is $S - C_1$. (The subtraction here is an oriented union, with the orientation of C_1 switched). Thus,

$$\int_{S - C_1} \omega = \int_D d\omega = 0.$$

If B is unorientable, we can use this same argument with the double covers of S and C_1. ∎

Section 4.4: Higher codimension

In higher codimension, there are two difficulties. First, if Theorem 1.2.1 fails, it may be because the normal wedges, though defined at each point of B, intersect each other. An example of this is a pair of k-planes in $\mathbf{R}^{2k}$, intersecting only at the origin. Since the curvature is zero, such a pair fails the curvature criterion only if the resulting normal wedges intersect. Unfortunately, it is easy for a pair of planes to be area-minimizing and yet be too close together for the normal wedges to be disjoint (see Chapter 5, Section 3).

The second difficulty, which we will deal with in more detail than the first, is that there may not exist a conservative unit normal vectorfield on B. Thus, when we construct a comparison surface S (induced by some nonconservative vectorfield), it will not be calibrated by the simple form ω as before. The normal part of the variation $\nabla_x \nu$ will give S more area. Since we are only accounting for part of the area of any comparison surface S, we cannot use this method as it stands, to prove that C is *not* area-minimizing.

We will first state the general theorem of this section, then find families of cones for which the hypotheses hold.

4.4.1 Theorem: Let B be a submanifold of $\mathbf{S}^{n-1}$, and $C = 0 \divideontimes B$. Suppose that there exists a conservative normal vectorfield $\nu(p)$ for $p \in B$ (cf. Definition 4.2.2). If for all $p \in B$ it is true that there is no solution of the initial value problem

$$(1) \qquad (g - \frac{t}{k}g')^2 + (\frac{g'}{k})^2 \leq \det(I - t\,\mathbf{h}_{ij}^{\nu(p)})^2, \qquad g(0) = 1$$

which reaches $g = 0$, then C is not area-minimizing.

Proof: The proof is the same as that of Theorem 4.3.1. We define a certain vectorfield on the normal wedge at p, get a planar curve γ, and let S be the surface induced by γ and $\nu(p)$; S will have less area than C_1.

One family of cones for which there *is* a conservative normal vectorfield are the cones over principal orbits of "polar" group actions.

4.4.2 Definition: Suppose that $B^{k-1} \subset \mathbf{S}^{n-1}$ is a principal orbit (i.e., an orbit of maximal dimension) of a group action on $\mathbf{R}^n$. Let $p \in B$, and let $\xi \in \mathbf{R}^n$ be a small $n - k$ dimensional disk centered at p, and normal to B. If all orbits near B intersect ξ orthogonally, then the group action is called "polar."

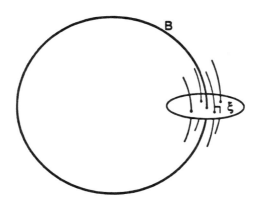

4.4.3 Lemma: If $B \subset \mathbf{S}^{n-1}$ is a principal orbit of a polar group action $G \subset SO(n)$, then we can choose a conservative unit normal vectorfield ν on B. In fact, if p is a (single) given point of B, we can arbitrarily specify the vector $\nu(p)$.

Proof: Pick $p \in B$, and choose an arbitrary (nearby) point z in the normal disk ξ. Let $\nu(p) = z - p$. If $g \in G$, let $\nu(gp) = gz - gp$. Since g induces an isometry of $\mathbf{R}^n$, ν is normal to B. Now if $\alpha(t)$ is a curve in G with $\alpha(0) = e$ (the identity), and if

$$\frac{d}{dt}(\alpha p) = X \in T_p(B),$$

then

$$\nabla_X \nu = \frac{d}{dt}(\alpha z - \alpha p)\big|_{t=0}.$$

Since $\frac{d}{dt}(\alpha z)$ and $\frac{d}{dt}(\alpha p)$ are both normal to ξ, they are both tangent (or parallel) to B, as is their difference. Since $\nu(p)$ has constant length, we can normalize it and get a conservative unit normal vectorfield. ∎

The next lemma states that if we have a collection $\{B_i\}$ of submanifolds of spheres, each having a conservative normal vectorfield, then their Cartesian product also has a conservative normal vectorfield.

4.4.4 Lemma: Let $\mathbf{S}^{n-1}(r)$ denote the sphere of radius r in $\mathbf{R}^n$, centered at 0. Let $B_i \subset \mathbf{S}^{n_i-1}(r_i)$, $1 \leq i \leq m$, with $\sum r_i^2 = 1$. Let

$$B = B_1 \times \cdots B_m \subset \mathbf{S}^{N-1}.$$

Suppose that each B_i is such that for any $q \in B_i$ and any $\nu \in T_q(\mathbf{S}^{n_i-1})$ which is normal to B_i at q, we can find a conservative normal vectorfield $\nu_i(p) \in T_p(\mathbf{S}^{n_i-1})$ such that $\nu(q) = \nu$. This (vacuously) includes the possibility that $B_i = \mathbf{S}^{n_i-1}$ for some values of i. Then the same holds true for B; if $q \in B$ and $\nu \in T_q(\mathbf{S}^{N-1})$ is normal to B, then we can find a conservative normal vectorfield $\nu(p)$ such that $\nu(q) = \nu$.

Proof: Let $\nu \in T_q(\mathbf{S}^{N-1})$ be normal to B. Let π_i denote orthogonal projection onto $\mathbf{R}^{n_i}$. Let $\nu_i = \pi_i(\nu)$, and $q_i = \pi_i(q)$. Then ν_i is normal to B_i, but may contain a radial component; write

$$\nu_i = \mu_i + \alpha_i \frac{\partial}{\partial r_i}.$$

(If $B_i = \mathbf{S}^{n_i-1}$ then $\mu_i = 0$). Since $\nu \in T_q(\mathbf{S}^{N-1})$, $\sum \alpha_i r_i = 0$. Now let

$$\nu_i(p_i) = \mu_i(p_i) + \alpha_i \frac{\partial}{\partial r_i}\Big|_{p_i}$$

for $p_i \in B_i$, where $\mu_i(p_i)$ is a conservative normal vectorfield agreeing at q_i with the (already specified) vector μ_i. Let $\nu(p)$ be the vectorfield whose projections onto $\mathbf{R}^{n_i}$ are $\nu_i(p_i)$. Then $\nu(p)$ is normal to B. Also, since $\sum \alpha_i r_i = 0$, $\nu(p) \in T_p(\mathbf{S}^{N-1})$.

We need to know that the covariant derivative of $\nu(p)$ (in all tangent directions) is tangent to B. It will be tangent to B if its projections onto $\mathbf{R}^{n_i}$ are tangent to B_i. The tangent vectors to B_i (when included into $\mathbf{R}^N$) form a basis for the tangent space to B, so we need to know that $\nabla_{X_i} \nu_i$ is tangent to B_i, for X_i tangent to B_i. But

$$\nabla_{X_i} \nu_i = \nabla_{X_i} \mu_i + \alpha_i \nabla_{X_i} \left(\frac{\partial}{\partial r_i}\right)$$

$$= \nabla_{X_i} \mu_i + \alpha_i X_i \in T_p(B_i). \quad \blacksquare$$

4.4.5 Theorem: Let B be a minimal submanifold of $\mathbf{S}^{n-1}$, of the form

$$B = B_1 \times \cdots \times B_m,$$

where each B_i lies in a (nonunit) sphere $\mathbf{S}^{n_i-1} \subset \mathbf{R}^{n_i} \subset \mathbf{R}^N$. Suppose that each B_i is an orbit of a Lie group with polar action. (This includes the possibility that $B_i = \mathbf{S}^{n_i-1}$ itself.) Suppose that Theorem 1.2.1 fails, for the reason that at some $q \in B$, the O.D.E. in 2.3.8 corresponding to some normal ν to C at q has no solution curve $g(t)$ connecting the point $(0, 1)$ to the t axis. Then the cone $C = 0 \times B$ is not area-minimizing. ∎

Proof: Since each B_i is isoparametric, so is B. In fact, there is an isometry of $\mathbf{R}^N$ taking any $p \in B$ to any $q \in B$, under which ν is invariant. Thus, the second fundamental form, when projected onto $\nu(p)$, gives the same matrix (or an orthogonally similar matrix) for all $p \in B$. Therefore, if Theorem 1.2.1 fails for the reason stated above, then we can choose a conservative normal vectorfield such that at each point $p \in B$, the O.D.E. in 2.3.8 corresponding to the designated normal $\nu(p)$ fails to have a solution curve satisfying $g(0) = 1$ and reaching $g = 0$. Then by Theorem 4.4.1, C is not area-minimizing.

4.4.6 Corollary: Let C be a cone as described in Theorem 4.3.1, Theorem 4.4.1, or Theorem 4.4.5. Let α be the sup norm of the second fundamental form, as defined in 1.3.3. If $\alpha > \frac{k-2}{2}$ then C is unstable.

Proof: Recall that $\alpha^2 = 8p_2$, where p_2 is the second-order coefficient of $p(t) = \det(I - t\,\mathbf{h}_{ij}^\nu)$. By Theorem 3.3.2, for *arbitrarily small* t_0 we can find a solution curve $g(t)$ of the differential equation (1) such that $g(t_0)$ lies on $\beta(t)$, and such that $g(0) < 1$. Such a curve g gives rise to a surface which is an arbitrarily small perturbation of the truncated cone, and which has less area. (If we wish to have S agree with C on a neighborhood of the origin, we can choose a tiny perturbation of the integral curve used in constructing S, and still have $\text{Area}(S') < \text{Area}(C_1)$.) ∎

4.4.7 Corollary: Let F be a family of cones over Cartesian products of B_i as described in Theorem 4.4.5. If each of the cones has normal radius at least twice as large as the focal radius, then Theorem 1.2.1 gives a necessary and sufficient condition for cones in this family to be area-minimizing.

Proof: Theorem 1.2.1 cannot fail for the reason that resulting normal wedges intersect; the normal wedges would automatically have radius smaller than the focal radius, and this would contradict Lemma 1.3.1. Thus, if Theorem 1.2.1 fails to prove that one of the cones of the family is area-minimizing, it fails for the reason that solution curves of the initial value problem in 2.3.8 do not reach $g = 0$.

Chapter 5: Examples of Area-minimizing Cones

Section 5.1: Cones over products of spheres

The first known area-minimizing hypersurface with an interior singularity was the cone over $\mathbf{S}^k \times \mathbf{S}^k$ for $k \geq 3$, discovered by Bombieri, DeGiorgi, and Giusti. Since then, other area-minimizing cones over products of spheres have been found (see the introduction to this paper). The following theorem completes the classification.

5.1.1 Theorem: Let C be a minimal cone over a product of two or more spheres. Then

(1): If $\dim(C) > 7$, C is area-minimizing;

(2): If $\dim(C) < 7$, C is unstable (and thus not minimizing);

(3): If $\dim(C) = 7$, then C is stable, and C is minimizing if and only if none of the spheres is a circle.

Proof: We will apply the criteria developed in Chapters 1 and 4. The first thing to do is to calculate the normal radius of C, and the second fundamental form. Once we have done this, we will see that for $\dim(C) > 7$, we can use the simpler form (Theorem 1.3.5) to prove that the cone is minimizing. When the dimension is less than 7, we will see that by Corollary 4.4.6, the cone is unstable, and thus not minimizing. If $\dim(C) = 7$, we will need to use Theorem 1.2.1, the stronger form of the minimization test.

Second fundamental form

Let $C = 0 \divideontimes B = 0 \divideontimes (\mathbf{S}^{a_1} \times \cdots \times \mathbf{S}^{a_m})$. For convenience, let

$$a_1 \leq a_2 \leq \cdots \leq a_m.$$

Let $\mathbf{S}^{a_i}$ have radius r_i, and $\sum r_i^2 = 1$. Since C is to be minimal, this determines the values of r_i, as we will see below. At the point

$$p = (r_1, 0, \ldots, 0, r_2, 0, \ldots, 0, \ldots, r_m, 0, \ldots, 0) \in C,$$

we will set up local coordinates on B:

$$(x_{11}, \ldots, x_{1\,a_1}, \ldots, x_{m1}, \ldots, x_{m\,a_m}) \rightarrow$$

$$\left(\frac{r_1}{N_1}, \frac{x_{11}}{N_1}, \ldots, \frac{x_{1\,a_1}}{N_1}, \ldots, \frac{r_m}{N_m}, \frac{x_{m\,1}}{N_m}, \ldots, \frac{x_{m\,a_m}}{N_m}\right)$$

where

$$N_i = \frac{1}{r_i}\sqrt{r_i^2 + \sum_j x_{ij}^2},$$

so that $(0, \ldots, 0) \to p$.

Then $\nabla_{e_i} e_j = 0$, for $i \neq j$, and for any j and k, if $e_i = \frac{\partial}{\partial x_{kj}}$ then

$$\nabla_{e_i} e_i = (0, \ldots, 0, \frac{1}{r_k}, 0, \ldots, 0),$$

where $\frac{1}{r_k}$ is in the position of r_k.

At p, the normal directions to C are all vectors of the form

$$\nu_\beta = (\beta_1, 0, \ldots, 0, \ldots, \beta_m, 0, \ldots, 0)$$

such that $\sum \beta_i r_i = 0$ and $\sum \beta_i^2 = 1$. Then $\mathbf{h}_{ij}^{\nu_\beta}$ is the diagonal matrix

$$\operatorname{diag}\left(\frac{\beta_1}{r_1}, \ldots, \frac{\beta_1}{r_1}, \ldots, \frac{\beta_m}{r_m}, \ldots, \frac{\beta_m}{r_m}\right)$$

where β_i/r_i appears a_i times;

$$\|\mathbf{h}_{ij}^{\nu_\beta}\| = \sum_i \left(a_i \frac{\beta_i^2}{r_i^2}\right)$$

and

$$\det(I - t\,\mathbf{h}_{ij}^{\nu_\beta}) = \prod_i (1 - \frac{\beta_i}{r_i}t)^{a_i}.$$

Since we want B to be minimal, the first order term of $\det(I - t\,\mathbf{h}_{ij}^{\nu_\beta})$ must vanish; thus, $\sum(a_i \frac{\beta_i}{r_i}) = 0$ for all $(\beta_1, \ldots, \beta_m)$ satisfying $\sum(\beta_i r_i) = 0$. For any i, j we can set

$$\beta = (0, \ldots, 0, r_j, 0, \ldots, 0, -r_i, 0, \ldots, 0)$$

with r_j in the ith position and r_i in the jth spot, thus giving

$$\frac{a_i}{r_i^2} = \frac{a_j}{r_j^2}.$$

Since we also want $\sum r_i^2 = 1$, we have

$$r_i = \frac{\sqrt{a_i}}{(\sum a_j)^{1/2}}.$$

Note that $r_1 \leq \cdots \leq r_m$. Now

$$\|\mathbf{h}_{ij}^{\nu_\beta}\| = ((\sum a_j) \sum \beta_i^2)^{1/2} = (\sum a_j)^{1/2}$$

$$\|\mathbf{h}_{ij}^{\nu_\beta}\|^2 = \sum a_j = \dim(B) = \dim(C) - 1.$$

This value will be used in Table 1.4.1 to find the vanishing angle of C. We will need to compare the vanishing angle with the normal radius, which we compute as follows.

Normal Radius

By rotational symmetry, the normal radius is the same at every point. Let us calculate it at

$$p = (r_1, 0, \ldots, 0, r_2, 0, \ldots, 0, \ldots, r_m, 0, \ldots, 0).$$

We wish to find the shortest normal geodesic in $\mathbf{S}^{n-1}$ which intersects C at some

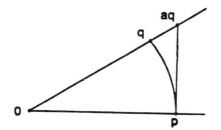

other point $q \neq p$. Equivalently, (as long as the normal radius θ is less than 90°) we can extend straight line segments normal to C at p, and minimize the norm a of the points where the segments intersect C.

Thus, we take a normal

$$\beta = (\beta_1, 0, \ldots, 0, \ldots, \beta_m, 0, \ldots, 0)$$

and add it to p to get another point in C:

$$(r_1 + \beta_1, 0, \ldots, r_m + \beta_m, 0, \ldots, 0) \in C;$$

$$r_1 + \beta_1 = \pm a r_1$$

$$r_2 + \beta_2 = \pm a r_2$$

$$\vdots$$

$$r_m + \beta_m = \pm a r_m$$

$$r_i = \frac{\beta_i}{1 + \sigma_i a}, \qquad \sigma_i = \pm 1.$$

Since $\beta \neq 0$, $a > 1$. Also, $\sum \beta_i r_i = 0$, thus

$$\sum (1 + \sigma_i a) r_i^2 = 0$$

$$\left(\sum r^2 \right) + a \sum \sigma_i r_i^2 = 0$$

$$1 < a = \frac{-1}{\sum \sigma_i r_i^2}.$$

Since $a > 1$, at least one σ_i is positive. Minimizing a under these constraints amounts to letting $\sigma_1 = 1$ (since r_1 is the smallest among the r_i), and $\sigma_i = -1$ for $i > 1$. Then the normal radius is

$$\theta = \cos^{-1} \left(\frac{1}{a} \right) = \cos^{-1} (-r_1^2 + r_2^2 + r_3^2 + \cdots + r_m^2)$$

$$= \cos^{-1} \left(\frac{-a_1 + a_2 + \cdots + a_m}{\sum a_i} \right) = \cos^{-1} \left(1 - \frac{2a_1}{k-1} \right).$$

Results for $k \neq 7$

Now we can apply Theorem 1.3.5. Using $\alpha^2 = k - 1$ in Table 1.4.1, we see that if $k < 7$, the vanishing angle does not exist, and in fact, (by Corollary 4.4.6), C is unstable. If $7 < k \leq 12$, the vanishing angle does exist, and is less than half of the normal radius; thus, C is area-minimizing. For example, the computation in dimension 12 is as follows:

$$k = 12$$

$$\alpha^2 = k - 1 = 11$$

$$\text{Vanishing angle} = \theta_0(p) \leq 8.88°$$

$$\text{Normal radius} = \cos^{-1} \left(1 - \frac{2a_1}{k-1} \right) \geq \cos^{-1} \left(1 - \frac{2}{11} \right) > 35° > 2(8.88°).$$

By Proposition 1.4.2, we can deduce that if $k > 12$ then

$$\tan(\theta_0(p)) \leq \frac{12}{k}\tan(8.88°) < \frac{2}{k}$$

whereas the tangent of half the normal radius is

$$\tan(\frac{1}{2}\cos^{-1}(1 - \frac{2}{k-1}))$$

$$= \frac{1}{\sqrt{k-2}} > \frac{2}{k}.$$

Dimension 7

If the dimension of C is 7, however, a solution of the initial value problem in Definition 1.1.6 exists for small t, but does not lead to a vanishing angle. Thus, Theorems 1.3.5 only tells us that C is stable. We need to minimize the quantity $\det(I - t\,\mathbf{h}_{ij}^\nu)$ by hand, instead of by Corollary 1.3.3 or Corollary 1.3.4. This is not too hard to do, with the machinery we have developed:

$$\det(I - t\,\mathbf{h}_{ij}^\nu) = \prod_i (1 - t(\frac{(\sum a_j)^{1/2}}{\sqrt{a_i}}\beta_i))^{a_i} = \prod(1 - \phi_i t)^{a_i}$$

where

$$\phi_i = \frac{(\sum a_j)^{1/2}}{\sqrt{a_i}}\beta_i.$$

Recall that we carefully defined the radii r_i in order that the first order term of the above determinant should vanish; thus, we know that $\sum(a_i\phi_i) = 0$. In order to minimize the determinant we can use Corollary A2, if we first make a substitution; let

$$\psi_1 = \cdots = \psi_{a_1} = \phi_1$$

$$\psi_{a_1+1} = \cdots = \psi_{a_1+a_2} = \phi_2$$

$$\text{etc.}$$

Then $\sum \psi_j = 0$ and $\sum \psi_j^2 = \sum a_j$;

$$\det(I - t\,\mathbf{h}_{ij}^\nu) = \prod(1 - \psi_i t)$$

$$\geq \left(1 - t\sqrt{\sum a_j}\sqrt{\frac{\dim(B)-1}{\dim(B)}}\right)\left(1 + t\sqrt{\sum a_j}\sqrt{\frac{1}{\dim(B)(\dim(B)-1)}}\right)^{\dim(B)-1}$$

$$= (1 - t\sqrt{5})(1 + \frac{t}{\sqrt{5}})^5.$$

This minimum is achieved by letting one of the ψ_i be positive, and the rest be negative (see Corollary A2). However, the ψ_i are equal in groups. For example, if one of the spheres defining the cone is $\mathbf{S}^3$, then three of the ψ_i must be equal. Since the above minimum depends on having one of the ψ_i different from the rest, we can only achieve it if one of the spheres is a circle ($\mathbf{S}^1$). In this case, the determinant equals $F(\alpha, t, m)$ as defined in Lemma 1.3.2, with $m = k-1 = 6$ and $\alpha = \sqrt{6}$. Since this is the function which was used to calculate Table 1.4.1, and since the table lists asterisks for $k = 7$ and $\alpha^2 = 6$, the vanishing angle does not exist in this case. Then by Theorem 1.2.1 and Corollary 4.4.7, the cones over $\mathbf{S}^1 \times \mathbf{S}^5$, $\mathbf{S}^1 \times \mathbf{S}^2 \times \mathbf{S}^3$, $\mathbf{S}^1 \times \mathbf{S}^1 \times \mathbf{S}^4$, etc., are stable but not minimizing.

The remaining cones in question are the cones over $\mathbf{S}^2 \times \mathbf{S}^4$, $\mathbf{S}^3 \times \mathbf{S}^3$, and $\mathbf{S}^2 \times \mathbf{S}^2 \times \mathbf{S}^2$. Since the first two are codimension 1 cones, there are only two normals. This makes the minimization of the determinant easier; for $\mathbf{S}^2 \times \mathbf{S}^4$ we have

$$\min \det(I - t\,\mathbf{h}^{\nu}_{ij}) = (1 + t\sqrt{2})^2 (1 - \frac{t}{\sqrt{2}})^4$$

and for $\mathbf{S}^3 \times \mathbf{S}^3$ we have

$$\min \det(I - t\,\mathbf{h}^{\nu}_{ij}) = (1 + t)^3 (1 - t)^3.$$

Using these values in Definition 1.1.6 or Theorem 2.3.6 and computing by numerical methods (see Section 3.4), we can confirm the previously known results that these two cones are minimizing. Incidentally, the vanishing angle for the former is about $19.9°$, and for the latter, $19°$. For $\mathbf{S}^2 \times \mathbf{S}^2 \times \mathbf{S}^2$, note that

$$\prod (1 - \phi_i t)^{a_i} = \prod (1 - \phi_i t)^2 = \left(\prod (1 - \phi_i t) \right)^2.$$

Minimizing this is equivalent to minimizing

$$\prod (1 - \phi_i t),$$

subject to $\sum \phi_i = 0$ and $\sum \phi_i^2 = 6/2 = 3$. By Corollary A2, this minimum is given by $\phi_1 = \sqrt{2}$ and $\phi_2 = \phi_3 = \frac{-1}{\sqrt{2}}$, so that

$$\min \det(I - t\,\mathbf{h}^{\nu}_{ij}) = (1 + t\sqrt{2})^2 (1 - \frac{t}{\sqrt{2}})^4;$$

thus, the calculation is the same as for $\mathbf{S}^2 \times \mathbf{S}^4$. ∎

Section 5.2: Cones over unorientable manifolds

Next we will prove that for $k \geq 4$, the cone over a certain embedding of a projective $k-1$ plane is minimizing modulo 2 (i.e., minimizing among unoriented surfaces; see [M6] and [F1, 4.2.26] for definitions). When k is odd, these are unorientable minimizing cones, the first known examples of such. Further examples have now been found; see [P]. If $k = 3$, it is not known whether the cone is minimizing.

Let B be the image in $\mathbf{R}^{k \times k} \cong \mathbf{R}^{k^2}$ of the unit $k-1$ sphere in $\mathbf{R}^k$, under the map

$$F : \mathbf{x} \to \mathbf{x}\,\mathbf{x}^T,$$

where $\mathbf{x}$ is viewed as a column vector. We will think of $\mathbf{x}\,\mathbf{x}^T$ interchangeably as a k by k matrix, or as an element of $\mathbf{R}^{k^2}$ defined in the natural way. In particular, the metric is the Euclidean metric on $\mathbf{R}^{k^2}$. Note that in matrix notation, the norm is given by

$$\|A\| = \mathrm{tr}(A\,A^T).$$

Thus, B lies in the unit sphere in $\mathbf{R}^{k^2}$, because

$$\|\mathbf{x}\,\mathbf{x}^T\| = \mathrm{tr}(\mathbf{x}\,\mathbf{x}^T\,\mathbf{x}\,\mathbf{x}^T) = \mathrm{tr}(\mathbf{x}\,(\mathbf{x}^T\,\mathbf{x})\,\mathbf{x}^T) = \mathrm{tr}(\mathbf{x}\,\mathbf{x}^T) = \mathbf{x}^T\,\mathbf{x} = 1.$$

The submanifold B looks the same at every point, so we need only apply the curvature criterion at a single point. To see this, let $F(\mathbf{x})$ and $F(\mathbf{y})$ be two points in B. We will define a linear isometry (rotation) Θ of $\mathbf{R}^{k^2}$ which leaves B invariant, and which takes $F(\mathbf{x})$ to $F(\mathbf{y})$. Namely, let Q be an orthogonal matrix such that $Q\mathbf{x} = \mathbf{y}$. Let $\Theta(A) = Q\,A\,Q^T$. This is a linear isometry on $\mathbf{R}^{k^2}$. Its restriction to B is induced by the map $\theta : \mathbf{x} \to Q\mathbf{x}$, since

$$F(Q\mathbf{x}) = (Q\mathbf{x})(Q\mathbf{x})^T = Q\mathbf{x}\mathbf{x}^T Q^T = \Theta(F(\mathbf{x})).$$

Since θ leaves $\mathbf{S}^{k-1}$ invariant, Θ leaves B invariant.

There is a small technicality which we will mention at this point. Since $\mathrm{tr}(F(\mathbf{x})) = x_1^2 + \cdots + x_k^2 = 1$, B lies in an affine hyperplane. If we took the cone over B with vertex at the origin, the cone would not be minimal. The center of B is at the point (or matrix) $D = \frac{1}{k}I$. Thus, let us define $F_1(\mathbf{x}) = F(\mathbf{x}) - D$, and call its image B_1. The manifold B_1 is centered at the origin, and is a minimal submanifold of the sphere of radius $\sqrt{(k-1)/k}$. For notational convenience, rather than dilating this so that it lies in the unit sphere, we will proceed with our computations at points of B_1, and compensate for this at the end (the curvature will be larger because we are closer to the origin.)

Notice that the above statements about symmetry hold for B_1 as well as B.

We will work with B_1 as it sits in $\mathbf{R}^{k^2}$, for convenience. However, we could reduce the codimension by about half; since each point (i.e., matrix) of B_1 is symmetric and has trace zero, B_1 can be rotated so that it sits in $\mathbf{R}^{k(k+1)/2 - 1}$.

Tangent and normal vectors

We set up local coordinates at the point $P = F_1(1, 0, \ldots, 0)$. The coordinates are induced by coordinates on $\mathbf{S}^{k-1}$:

$$(y_1, \ldots, y_{k-1}) \rightarrow \frac{(1, y_1, \ldots, y_{k-1})}{\|(1, y_1, \ldots, y_{k-1})\|} \rightarrow$$
$$\frac{1}{1 + y_1^2 + \cdots + y_{k-1}^2} \left((1, y_1, \ldots, y_{k-1})(1, y_1, \ldots, y_{k-1})^T\right).$$

The associated basis of tangent vectors $f_i (1 \leq i \leq k-1)$ at P is then

$$f_i = e_1 e_{i+1}^T + e_{i+1} e_1^T,$$

where e_i is the unit (column) k-vector with a 1 in the ith slot.

The normals at $F_1(e_1)$ to the cone $C = 0 \ast B_1$ are linear combinations of

$$\nu_i = e_1 e_1^T + (k-1)e_{i+1} e_{i+1}^T, \qquad (1 \leq i \leq k-1),$$

$$\mu_{1,i} = e_1 e_{i+1}^T - e_{i+1} e_1^T, \qquad (1 \leq i \leq k-1),$$

$$\mu_{i,j} = e_{i+1} e_{j+1}^T, \qquad (2 \leq i, j \leq k-1 \quad \text{and} \quad i \neq j).$$

Normal radius

5.2.1 Lemma: The normal radius of C is at least $90°$; thus, we can allow vanishing angles of at least $45°$.

Proof: We show that $F_1(e_1)$ plus a vector normal to C at $F_1(e_1)$ can never lie in C. To see this, suppose that $\mathbf{x} \neq e_1$ and $a > 0$, and

$$aF_1(\mathbf{x}) = a(\mathbf{x}\,\mathbf{x}^T - \frac{1}{k}I) = e_1 e_1^T - \frac{1}{k}I + \nu,$$

for some normal vector ν. First notice that the left-hand side is a symmetric matrix, so the normal is symmetric as well; looking at the above list of normals, we see that any symmetric normal at $F_1(e_1)$ has no off-diagonal component in the first row or column. Thus, the same is true of $\mathbf{x}\mathbf{x}^T$, so that $x_1 x_i = 0$ for $1 < i \leq k$. If x_1 were nonzero, then all of the other x_i would be zero, so that $\mathbf{x} = e_1$, a contradiction. Therefore, $x_1 = 0$. Also notice that by taking the trace of both sides of the above matrix equation, we see that $\operatorname{tr}(\nu) = 0$. Now if α_i are the coefficients of ν_i in the expansion of ν in terms of the basis of normals given above, then $\operatorname{tr}(\nu) = k \sum \alpha_i$, so that $\sum \alpha_i = 0$. But the 1,1 entry of $e_1 e_1^T - \frac{1}{k} I + \nu$ is then $1 - 1/k + \sum \alpha_i = 1 - 1/k > 0$, while the 1,1 entry of $a(\mathbf{x}\mathbf{x}^T - \frac{1}{k} I)$ is $a(x_1^2 - 1/k) = a(-1/k) < 0$, a contradiction. $\blacksquare$

Second fundamental form

We now need to compute the second fundamental form, project it onto an arbitrary unit normal, and maximize the Euclidean norm of the resulting matrix. We find that

$$\nabla_{f_i}(f_i) = -2e_1 \, e_1^T + 2e_{i+1} \, e_{i+1}^T$$
$$\nabla_{f_i}(f_j) = e_{i+1} \, e_{j+1}^T + e_{j+1} \, e_{i+1}^T \qquad \text{if} \quad i \neq j.$$

The projection of any of these onto the normal $\mu_{1,i}$ is zero for all i, so we may as well take our unit normal to be

$$\nu = \sum(\alpha_i \nu_i) + \sum_{1 < j, l \leq k-1} (\beta_{j,l} \, \mu_{j,l}),$$

whose norm is $(\sum \alpha_i)^2 + (k-1)^2 \sum(\alpha_i^2) + \sum \beta_{j,l}^2$, which we require to be 1. The second fundamental form h' corresponding to this normal direction is the matrix

$$\begin{pmatrix} 2(k-1)\alpha_1 - 2\sum \alpha_i & \beta_{12} + \beta_{21} & \cdots & \beta_{1,k-1} + \beta_{k-1,1} \\ \beta_{12} + \beta_{21} & 2(k-1)\alpha_2 - 2\sum \alpha_i & \cdots & \beta_{2,k-1} + \beta_{k-1,2} \\ \vdots & \vdots & \ddots & \vdots \\ \beta_{1,k-1} + \beta_{k-1,1} & \beta_{2,k-1} + \beta_{k-1,2} & \cdots & 2(k-1)\alpha_{k-1} - 2\sum \alpha_i \end{pmatrix}.$$

(We have written h' instead of h because we will need to normalize this; see the corrections below). Then

$$\|h'\|^2 = -4(k-1)(\sum \alpha_i)^2 + 4(k-1)^2 \sum(\alpha_i^2) + 2 \sum_{j < l} (\beta_{j,l} + \beta_{l,j})^2;$$

since

$$4(\sum \alpha_i)^2 + 4(k-1)^2 \sum (\alpha_i^2) + 4 \sum \beta_{j,l}^2 - 4 = 0,$$

we can subtract this to get

$$\|h'\|^2 = 4 - 4k(\sum \alpha_i)^2 + 2 \sum_{j<l} (\beta_{j,l} + \beta_{l,j})^2 - 4 \sum \beta_{j,l}^2$$

$$= 4 - 4k(\sum \alpha_i)^2 - 2 \sum_{j<l} (\beta_{j,l} - \beta_{l,j})^2 \leq 4,$$

which we maximize (for example) by setting $\alpha_i = 0$, $\beta_{j,l} = \beta_{l,j}$.

There are two corrections to make at this point; we want the maximum of $\|h\|$ with respect to *orthonormal* tangent vectors, and evaluated at points of the unit sphere. Since $\|f_i\| = \sqrt{2}$, we divide $\|h'\|$ by $\sqrt{2}^2 = 2$. Since B_1 lies in the sphere of radius $\sqrt{(k-1)/k}$, we multiply $\|h'\|$ by this factor as well, giving

$$\|h\|^2 = \frac{k-1}{k}.$$

We can now use these computations to give us the following theorem.

5.2.2 Theorem: Let B_1 be the embedding of $\mathbf{RP}^{k-1}$ into $\mathbf{R}^{k^2}$ given as the image of the unit $k-1$ sphere under the map

$$F : \mathbf{x} \to \mathbf{x}\mathbf{x}^T - D,$$

where $\mathbf{x}$ is viewed as a column vector, and D is the centering factor $\frac{1}{k}I$. Let C be the cone over B_1. If $k \geq 4$, then C is area-minimizing (in the class of integral currents modulo 2).

Proof: If $k = 4$, by Table 1.4.1 the vanishing angle is $31.55°$, which by Lemma 5.2.1 is less than half the normal radius, so by Theorem 1.3.5, the cone is area-minimizing. Similarly, using Table 1.4.1 and Proposition 1.4.2, the cones over B_1, for all $k \geq 4$, are area-minimizing. ∎

If $k = 3$, the curvature criterion fails. It is not known whether the cone over this embedding of $\mathbf{RP}^2$ into $\mathbf{R}^9$ (or into $\mathbf{R}^5$; see comment on codimension at the first of this section) is area-minimizing modulo 2. However, Tim Murdoch [Mu] has used a beautiful new concept of calibrations to show that this cone is in fact area-minimizing among a large class of comparison surfaces.

Section 5.3: Unions of planes

The angle criterion (see [L1],[M7],[M8],[N]) tells exactly which pairs of oriented k-planes are area-minimizing. For many of these pairs, the curvature criterion Theorem 1.2.1 fails. For example, if two planes through the origin are very close together and have opposite orientations, then the pair is not minimizing, since a small strip spanning their boundaries will have less area. However, if they are close together and have the same orientation, the pair may still be minimizing. Since the curvature criterion does not deal with orientation, it will fail for such minimizing pairs. However, even if the angle criterion tells us that a pair of planes is minimizing with respect to any orientation, the curvature criterion *may still fail*. In certain "equal-angles" cases, as the dimension of the planes goes to infinity, the curvature criterion does asymptotically as well as the angle criterion.

It is conjectured that a pair of nonoriented planes through the origin minimizes area among nonoriented surfaces, if and only if the oriented pair is area-minimizing with respect to any orientation. The conjecture is true for 2-planes; see [M8].

Not much is known about exactly which m-tuples of k-planes through a common vertex are area-minimizing (see [M7]). Any m-tuple which is simultaneously calibrated will be minimizing; thus such calibrations as the special Lagrangian form give us many examples. This paper gives others (in fact, an open set of examples).

Let θ be the vanishing angle in dimension k with curvature zero (see Definition 1.1.7 and Table 1.4.1). Let the union of normal wedges of radius θ around a given k-plane be called a "θ-cushion." If the θ-cushions about all of the k-planes do not intersect, then the m-tuple will be area-minimizing, as well as minimizing modulo 2. See also [L2].

Section 5.4: Cones over compact matrix groups

In this section, we will prove that the cones over the matrix groups $O(m)$, $SO(m)$, $U(m)$, and $SU(m)$ are area-minimizing, for $m \geq 2$. These groups are thought of as sitting in $\mathbf{R}^{m^2}$ or in $\mathbf{C}^{m^2} = \mathbf{R}^{2m^2}$. Each is contained in the sphere of radius $\sqrt{m}$, centered at the origin. Benny Cheng [ChB] has already shown that the cone over $SU(m)$ is special Lagrangian (and thus area-minimizing), and has proved that the cone over $U(m)$ is area-minimizing for $m \geq 4$.

Computation of the second fundamental form

Because these are rotation groups, the second fundamental form is the same at each point. We will calculate it at I, after which we will make the necessary adjustment to find the second fundamental form at points of the *unit* sphere.

Notation: Let e_i be the m-vector with a 1 in the ith slot. Let $E_{ij} = e_i e_j^T$.

Let us write down the tangents and normals to each of the groups. For $SO(m)$ and $O(m)$, a basis of tangent vectors at the identity are the skew-symmetric matrices

$$X_{ij} = E_{ij} - E_{ji}, \quad 1 \leq i < j \leq m.$$

For $U(m)$, we also have the skew-Hermitian matrices

$$Y_{ij} = \mathbf{i}(E_{ij} + E_{ji}),$$

and

$$Z_j = \sqrt{2}\mathbf{i}E_{jj},$$

where the factor of $\sqrt{2}$ gives these vectors the same norm as X_{ij} and Y_{ij}. For $SU(m)$, we have the same tangent vectors, except that in place of the Z_j we need $Z_1 - Z_2$, $Z_1 - Z_3$, $Z_1 - Z_4$, etc., because the imaginary part of the trace has to be zero in this case.

For $SO(m)$ and $O(m)$, the normals at I are the symmetric matrices with trace zero (they are traceless since I is also a tangent vector, not mentioned above because the curvature in that direction is zero). For $U(m)$ the normals are the Hermitian matrices with trace zero. For $SU(m)$, there is an extra normal which is $\mathbf{i}I$.

We wish to calculate the second fundamental form on each of these groups, project it onto an arbitrary unit normal M, and maximize (over all such M) the Euclidean norm of the resulting matrix. Before proceeding, it will simplify things to note that we can assume M is a *diagonal* matrix. Further, except for $SU(m)$, the normal will be real and have trace zero. In the case of $SU(m)$, $\text{Im}(M) = \alpha I$ for some α. Let us prove this, first for $SO(m)$:

If M is a normal at I, then M is symmetric and has trace zero. There exists an orthogonal matrix Q such that QMQ^T is diagonal (and has trace zero). Since the transformation $X \to QXQ^T$ is an isometry preserving $SO(m)$ and its normal space (and leaving I fixed), it follows that the second fundamental form is the same when projected onto the normal M as onto the normal QMQ^T (after an orthonormal change of basis).

Similarly for $O(m)$. For $U(m)$, a normal M is Hermitian and has trace zero. There exists a unitary U such that UMU^* is real, diagonal, and traceless. The transformation $X \to UXU^*$ is an isometry of $\mathbf{C}^{m^2} \cong \mathbf{R}^{2m^2}$. For $SU(m)$, a normal is $M = \alpha \mathbf{i} I + A$, where A is Hermitian with trace zero. Then there exists U such that $UMU^* = \alpha \mathbf{i} I + \Lambda$, with Λ real, diagonal, and traceless.

Now we calculate the norm of the second fundamental form on $SO(m)$, projected onto an arbitrary (unit length) traceless diagonal matrix

$$A = \mathrm{diag}(a_1, \ldots, a_m).$$

Let

$$h_{ij} = \begin{pmatrix} \nabla_{X_{12}} X_{12} & \nabla_{X_{12}} X_{13} & \cdots & \nabla_{X_{12}} X_{m-1,m} \\ \nabla_{X_{13}} X_{12} & \nabla_{X_{13}} X_{13} & \cdots & \nabla_{X_{13}} X_{m-1,m} \\ \vdots & \vdots & \ddots & \vdots \\ \nabla_{X_{m-1,m}} X_{12} & \nabla_{X_{m-1,m}} X_{13} & \cdots & \nabla_{X_{m-1,m}} X_{m-1,m} \end{pmatrix}.$$

In a matrix Lie group, if X and Y are matrices representing tangents to the group at I, then

$$\nabla_X Y = \frac{1}{2}(XY + YX).$$

Using this equation, we can check that the only entries (matrices) in the above matrix h_{ij} having any nonzero diagonal elements are the entries

$$\nabla_{X_{ij}} X_{ij} = X_{ij}^2.$$

The list of these matrices X_{ij}^2 is

$$-E_{11} - E_{22}, -E_{11} - E_{33}, \ldots, -E_{m-1,m-1} - E_{mm}.$$

Projecting each onto A we get

$$h_{ij}^A = \begin{pmatrix} -a_1 - a_2 & 0 & \cdots & 0 \\ 0 & -a_1 - a_3 & \cdots & 0 \\ \vdots & \vdots & \ddots & \vdots \\ 0 & 0 & \cdots & -a_{m-1} - a_m \end{pmatrix}$$

so that

$$\|h_{ij}^A\|^2 = \sum_{i<j}(a_i + a_j)^2$$

$$= (m-1)\sum a_i^2 + \sum_{i<j} 2a_i a_j$$

$$= (m-2)\sum a_i^2 + \left(\sum a_i\right)^2$$

$$= (m-2)\sum a_i^2$$

$$= m-2.$$

Now we want this norm with respect to *unit* tangent vectors, at points of the *unit* sphere. Since the tangent vectors we have used have norm $\sqrt{2}$, we correct the squared norm by dividing by $(\sqrt{2}\,\sqrt{2})^2 = 4$. Since the norm of the matrix I is $\sqrt{m}$, we multiply the squared norm by $\sqrt{m}^2 = m$. The final result is

$$\|h_{ij}^A(\frac{1}{\sqrt{m}}I)\|^2 = \frac{m(m-2)}{4}.$$

The same calculations hold for $O(m)$. For $U(m)$, we also have the tangent vectors Y_i and Z_i to take into account. By inspection we can see that the only entries of the second fundamental form having nonzero projection onto the diagonal normal A are $\nabla_{X_i}X_i$, $\nabla_{Y_i}Y_i$, and $\nabla_{Z_i}Z_i$. (Note: $\nabla_{X_i}Y_i$ also has nonzero diagonal entries. However, they are imaginary. Thus, as an element of $\mathbf{R}^{2m^2}$, the projection of $\nabla_{X_i}Y_i$ onto a *real* diagonal matrix is zero. Further, its projection onto αiI, (which is one of the normals to $SU(m)$) is also zero, because $\operatorname{tr}(\nabla_{X_i}Y_i) = 0$.)

Then h_{ij}^A is the matrix

$$-\operatorname{diag}(a_1 + a_2, a_1 + a_3, \ldots, a_{m-1} + a_m,$$

$$a_1 + a_2, a_1 + a_3, \ldots, a_{m-1} + a_m, 2a_1, 2a_2, \ldots, 2a_m).$$

Using our previous calculation for $SO(m)$, we see that

$$\|h_{ij}^A\|^2 = 2(m-2) + 4\sum a_i^2 = 2(m-2) + 4 = 2m.$$

As with $SO(m)$ and $O(m)$, we compensate for the nonunit tangent vectors and the norm of I, and get

$$\|h_{ij}^A(\frac{1}{\sqrt{m}}I)\|^2 = \frac{m^2}{2}.$$

For $SU(m)$, the squared norm is less than $m^2/2$. To see this, first notice that none of the matrices $\nabla_{X_i} X_i$, $\nabla_{Y_i} Y_i$, $\nabla_{Z_i} Z_i$ have nonzero projection onto $\mathbf{i}I$. Thus, we may as well assume that the normal A is Hermitian, with trace zero, just as with $U(m)$. Now in this case we do not have the tangent vectors Z_i, but rather $Z_1 - Z_2$, $Z_2 - Z_3$, etc. Let us choose an orthonormal basis $\{W_i\}$ spanning the same space as $\{Z_i\}$, with

$$W_m = \frac{1}{\sqrt{m}}\mathbf{i}I,$$

so that for $i < m$, W_i is tangent to $SU(m)$. Now the projection of the matrix

$$\begin{pmatrix} \nabla_{W_1} W_1 & \cdots & \nabla_{W_1} W_m \\ \vdots & \ddots & \vdots \\ \nabla_{W_m} W_1 & \cdots & \nabla_{W_m} W_m \end{pmatrix}$$

onto A has the same Euclidean norm as the corresponding matrix $\langle \nabla_{Z_i} Z_j, A \rangle$. Taking away the last row and column (since W_m is not tangent to $SU(m)$) reduces the Euclidean norm. In the particular case of $SU(2)$, the norm is zero, since $SU(2)$ is just a round 3-sphere; the cone over it is an $\mathbf{R}^4$ sitting in $\mathbf{R}^8$. For $m > 2$, the upper bound $m^2/2$ will be good enough for our purposes.

Calculation of the normal radius

We now calculate the normal radius to C at I for each of these matrix groups, beginning with $O(m)$. We want to know the smallest tangent vector (matrix) T which we can add to I, and get a multiple of an orthogonal matrix. Since T is symmetric, $I + T$ is also symmetric. Another way to state the problem is that we need to find the symmetric orthogonal matrix R which is nearest to I in the Euclidean norm on $\mathbf{R}^{m^2}$. Since R is symmetric, there exists $Q \subset O(m)$ such that $QRQ^T = \Lambda$, a diagonal matrix. Notice that

$$\|R - I\| = \|Q(R - I)Q^T\| = \|\Lambda - I\|,$$

so we have reduced the problem to finding a diagonal orthogonal matrix nearest to I. Clearly, each entry of Λ is ± 1, so a nearest one is $\Lambda = \mathrm{diag}(-1, 1, 1, \ldots, 1)$. The angle between I and Λ is the normal radius of $O(m)$, which is

$$\cos^{-1}\left(\frac{I \cdot \Lambda}{\|I\| \, \|\Lambda\|}\right) = \cos^{-1}\left(\frac{\mathrm{tr}\Lambda}{m}\right) = \cos^{-1}\left(\frac{m-2}{m}\right).$$

To find the normal radius of $SO(m)$, the only difference is that an even number of the entries of Λ must be negative, so that a nearest Λ to I is

$$\Lambda = \operatorname{diag}(-1, -1, 1, 1, \ldots, 1).$$

The normal radius in this case is

$$\cos^{-1}\left(\frac{m-4}{m}\right).$$

A similar analysis goes through to show that the normal radius to $U(m)$ is

$$\cos^{-1}\left(\frac{m-2}{m}\right).$$

In the case of $SU(m)$, we might expect the normal radius to be $\cos^{-1}(\frac{m-4}{m})$. However, it is less than this because of the extra tangent vector $\mathbf{i}I$. It turns out that the cosine of the normal radius is $\frac{m-2}{m}\cos(\frac{\pi}{m-2})$. We will not use this precise value, but only outline a proof that the normal radius is greater than π/m. First, one shows that there will be a special unitary matrix nearest to I which is of the form $\Lambda = \operatorname{diag}(e^{i\theta_1}, \ldots, e^{i\theta_m})$ where $\sum \theta_j = 2\pi$, $0 \le \theta_j \le \pi$, and $\sin(\theta_i) = \sin(\theta_j)$ for each i, j. From this we can see that the three (essentially different) candidates for a nearest Λ are given by

$$\theta_1 = \theta_2 = \pi, \; \theta_3 = \cdots = \theta_m = 0,$$

$$\theta_1 = \pi - \frac{\pi}{m-2}, \; \theta_2 = \cdots = \theta_m = \frac{\pi}{m-2},$$

and

$$\theta_1 = \cdots = \theta_m = \frac{2\pi}{m}.$$

By taking dot products, we can see that the angles between these three candidates and the matrix I, are

$$\cos^{-1}\left(\frac{m-4}{m}\right), \; \cos^{-1}\left(\frac{m-2}{m}\cos(\frac{\pi}{m-2})\right), \text{ and } \frac{2\pi}{m}.$$

The second of these angles is the smallest, which can be proved with a bit of analysis. We care only that each angle is greater than π/m. To see (for example) that

$$\frac{m-2}{m}\cos(\frac{\pi}{m-2}) < \cos(\frac{\pi}{m}),$$

i.e., that

$$(m-2)\cos(\frac{\pi}{m-2}) < m\cos(\frac{\pi}{m}),$$

set $f(x) = x\cos(\frac{\pi}{x})$ and note that $f'(x) > 0$ on $(2, \infty)$. For reference below, we also note that

$$\cos^{-1}(\frac{m-4}{m}) > \cos^{-1}(\frac{m-2}{m}) > \frac{\pi}{m}, \quad \text{for } m \geq 4.$$

The second inequality is true because

$$\cos^2\frac{\pi}{m} = 1 - \sin^2\frac{\pi}{m} > 1 - \frac{\pi^2}{m^2} > 1 - \frac{4}{m} + \frac{4}{m^2} = (\frac{m-2}{m})^2.$$

We have now assembled the necessary information to prove that the cones in question are area-minimizing:

5.4.1 Theorem: The cones over the groups $SO(m)$, $O(m)$, $U(m)$, and $SU(m)$ are all area-minimizing, for $m \geq 2$. (These are the standard matrix representations, thought of as sitting in $\mathbf{R}^{m^2}$ and $\mathbf{C}^{m^2} \cong \mathbf{R}^{2m^2}$).

Proof: First we note three special cases:

(a) $0 \ast SO(2)$ is a 2-plane,

(b) $0 \ast O(2)$ is a pair of orthogonal planes in $\mathbf{R}^4$, and

(c) $0 \ast SU(2)$ is a 4-plane in $\mathbf{R}^8$.

To see that (c) is true, note that for arbitrary α and β, a complex matrix of the form

$$U = \begin{pmatrix} \alpha & \beta \\ -\overline{\beta} & \overline{\alpha} \end{pmatrix}$$

satisfies $UU^* = aI$, where $a = \alpha\overline{\alpha} + \beta\overline{\beta}$. Also, $\det U = a$. Thus, $\frac{1}{\sqrt{a}}U$ is special unitary, so that $U \in 0 \ast SU(2)$. The set of all matrices of the above form is a 4-plane in $\mathbf{R}^8$, spanned by the vectors (matrices) determined by $(\alpha, \beta) = (1,0), (\mathbf{i},0), (0,1), (0,\mathbf{i})$, respectively. This tells us that $C = 0 \ast SU(2)$ *contains* a flat 4-plane. However, since $SU(2)$ is an analytic, connected, three-dimensional submanifold of the (nonunit) 7-sphere, the cone C cannot contain anything more than this 4-plane.

It is well-known that the cones (a), (b), and (c) are area-minimizing. For the remaining cases, we use Theorem 1.3.5. We assemble the necessary information in a table, and note that the normal radius in each case is more than twice the vanishing angle, so that all of the cones are area-minimizing. ∎

Table 5.4.2: Data for matrix groups

Submanifold of sphere	Dimension of cone	$\|h_{ij}^M\|^2$	Vanishing angle	Normal radius
$SO(3)$	4	$3/4$	$< 32°$	$\cos^{-1}(-1/3) > 90°$
$O(3)$	4	$3/4$	$< 32°$	$\cos^{-1}(1/3) > 70°$
$SU(3)$	9	$< 9/2$	$< 12°$	$\cos^{-1}(-1/3) > 90°$
$U(2)$	5	2	$< 30°$	$90°$
$U(3)$	10	$9/2$	$< 11°$	$\cos^{-1}(1/3) > 70°$
$SO(m)$	$\binom{m}{2}+1$	$\frac{m(m-2)}{4}$	$< \frac{\pi}{2m}$	$\cos^{-1}(\frac{m-4}{m}) > \frac{\pi}{m}$
$O(m)$	$\binom{m}{2}+1$	$\frac{m(m-2)}{4}$	$< \frac{\pi}{2m}$	$\cos^{-1}(\frac{m-2}{m}) > \frac{\pi}{m}$
$SU(m)$	m^2	$< \frac{m^2}{2}$	$< \frac{\pi}{2m}$	$\cos^{-1}\left(\frac{m-2}{m}\cos(\frac{\pi}{m-2})\right) > \frac{\pi}{m}$
$U(m)$	m^2+1	$\frac{m^2}{2}$	$< \frac{\pi}{2m}$	$\cos^{-1}(\frac{m-2}{m}) > \frac{\pi}{m}$

The figures in the vanishing angle column are obtained from Table 1.4.1 and Proposition 1.4.2. The full details are not worth belaboring, in proving that the last four entries are less than $\frac{\pi}{2m}$; the main principle (by Proposition 1.4.2) is the following: Since the dimension is growing like m^2, we can let $\|h_{ij}\|^2$ grow as fast as m^4, whereas it actually only grows like m^2. The vanishing angles will then decrease on the order of $\frac{1}{m^2}$, and we are only requiring them to decrease on the order of $\frac{1}{m}$. ∎

Section 5.5: Codimension 1 cones over orbits of group actions

In this section we show which of the "homogeneous" codimension 1 cones are area-minimizing. A list of the cones appears in a table at the end of Lawson's paper [Ls]. We will refer to this list, and will use the volume functions given in the table for each cone.

Lawson proves in his paper that some of the homogeneous cones are area-minimizing, thus giving a partial classification. Others ([Hs], [Fm]) have worked out a complete list telling which are minimizing. The classification we give in this paper differs from previous results, in that we show that the cone over $\mathbf{S}^1 \times \mathbf{S}^5$ and the cone over $(SO(2) \times SO(8))/(\mathbf{Z}_2 \times SO(6))$ are actually stable but not area-minimizing. The former cone (see the Introduction and Section 5.1) was already shown by Simoes [Sm] not to be minimizing.

In the last column of Lawson's table we find V^2, the square of a certain volume function, for each of the homogeneous cones. We can use this to compute the Jacobian $p(t) = \det \left(I - t \, \mathbf{h}_{ij}^{\nu} \right)$.

Each cone can be described by taking a ray Q extending from the origin in a 2-plane P, embedding P as a subspace of $\mathbf{R}^n$, and letting a certain Lie group act on Q. Thus, the cone is the union of $n - 2$ dimensional orbits in $\mathbf{R}^n$, passing through points of Q.

The volume function $V(x, y)$ given by Lawson tells us the volume of an orbit passing through the point (x, y) in the plane P. Since the cone is to be minimal, Q

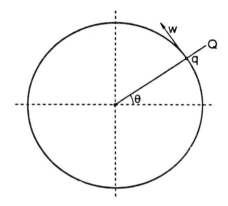

must pass through a point $q \in \mathbf{S}^1$ at which V is maximized over the unit circle.

If $\mathbf{w}$ is a tangent vector to $\mathbf{S}^1$ at q, then the Jacobian we need for Theorem 2.3.6 is

$$p(t) = \det \left(I - t \, \mathbf{h}_{ij}^{\nu} \right) = \frac{V(q + t\mathbf{w})}{V(q)}.$$

(Since we actually want $p(t) = \inf \det(I - t \, \mathbf{h}_{ij}^{\nu})$, we must compare $p(t)$ with $p(-t)$ and choose the appropriate one).

Thus, for each cone in Lawson's table we need to maximize V over $\mathbf{S}^1$, then rotate the coordinates and compute

$$\frac{V(1, t)}{V(1, 0)}.$$

As an example, we will do this computation for the family of cones in line (2) of the table.

By substituting $x = \cos\theta$ and $y = \sin\theta$ and differentiating, we find that the maximum of V on $\mathbf{S}^1$ occurs when $\tan^2(2\theta) = k - 2$. Setting

$$x = x' \cos\theta - y' \sin\theta$$

$$y = x' \sin\theta + y' \cos\theta,$$

we get

$$V^2(x,y) = (xy)^{2k-4}(x^2 - y^2)^2 =$$

$$\left(\frac{1}{2}(x'^2 - y'^2)\sin 2\theta + x'y'\cos 2\theta\right)^{2k-4}\left((x'^2 - y'^2)\cos 2\theta - 2x'y'\sin 2\theta\right)^2.$$

From the value of $\tan^2(2\theta)$ given above, we compute $\cos(2\theta) = \sqrt{\frac{1}{k-1}}$ and $\sin(2\theta) = \sqrt{\frac{k-2}{k-1}}$. Using these values, and setting $x' = 1$, $y' = t$, we get

$$V^2(x,y) = \left(\frac{1}{2}\sqrt{\frac{k-2}{k-1}}(1-t^2) + \frac{t}{\sqrt{k-1}}\right)^{2k-4}\left(\frac{1-t^2}{\sqrt{k-1}} - 2t\sqrt{\frac{k-2}{k-1}}\right)^2.$$

Table 5.5.1 gives the information we need in order to decide whether each of the cones in [Ls] is area-minimizing. The results are tabulated in 5.5.2. For most of the cones, we can use the simplified version of the curvature criterion, Theorem 1.3.5, together with Table 1.4.1, Proposition 1.4.2, Theorem 4.3.1, and Corollary 4.4.6. There are a few cones which this method only tells us are stable. These are the cones on the following lines of the table:

Line (1), $r + s = 8$,
Line (2), $8 \le m \le 11$,
Line (3), $4 \le m \le 5$,
Line (4), $m = 2$,
Line (5), and
Line (8).

These "borderline cases" we handle with Theorems 1.2.1 and 4.3.1, solving the corresponding differential equation numerically, using the actual function $p(t)$, rather than just the second-order coefficient. Of these twelve stable cones, two are not area-minimizing.

Table 5.5.1: Curvature calculations for codimension 1, homogeneous cones

Submanifold of sphere	Dimension of cone	$V'(1,t)/V'(1,0) = p(t)$	$\lvert 2p_2 \rvert = \alpha^2$
$\mathbf{S}^{r-1} \times \mathbf{S}^{s-1}$	$r+s-1$	$\left(1 - t\sqrt{\frac{r-1}{s-1}}\right)^{s-1}\left(1 + t\sqrt{\frac{s-1}{r-1}}\right)^{r-1}$	$r+s-2$
$\dfrac{SO(2) \times SO(m)}{\mathbf{Z}_2 \times SO(m-2)}$	$2m-1$	$(1 + \frac{2t}{\sqrt{m-2}} - t^2)^{m-2}(1 - 2t\sqrt{m-2} - t^2)$	$6m-6$
$\dfrac{SU(2) \times SU(m)}{T^1 \times SU(m-2)}$	$4m-1$	$(1 + \frac{2t}{\sqrt{m-3/2}} - t^2)^{2m-3}(1 - 2t\sqrt{m-\frac{3}{2}} - t^2)^2$	$12m-6$
$\dfrac{Sp(2) \times Sp(m)}{Sp(1)^2 \times Sp(m-2)}$	$8m-1$	$(1 + \frac{2t}{\sqrt{m-5/4}} - t^2)^{4m-5}(1 - 2t\sqrt{m-\frac{5}{4}} - t^2)^4$	$24m-6$
$\dfrac{U(5)}{SU(2) \times SU(2) \times T^1}$	19	$(1 + \frac{4t}{\sqrt{5}} - t^2)^5(1 - t\sqrt{5} - t^2)^4$	54
$\dfrac{SO(3)}{\mathbf{Z}_2 \times \mathbf{Z}_2}$	4	$1 - 3t^2$	6
$\dfrac{SU(3)}{T^2}$	7	$(1 - 3t^2)^2$	12
$\dfrac{Sp(3)}{Sp(1)^3}$	13	$(1 - 3t^2)^4$	24
$\dfrac{Sp(2)}{T^2}$	9	$(1 - 6t^2 + t^4)^2$	24
$\dfrac{G_2}{T^2}$	13	$(1 - 15t^2 + 15t^4 - t^6)^2$	60
$\dfrac{F_4}{Spin(8)}$	25	$(1 - 3t^2)^8$	48
$\dfrac{Spin(10) \times U(1)}{SU(4) \times T^1}$	31	$(1 - 3t^2)^{10}$	60

Theorem 5.5.2: Area-minimization results corresponding to Table 5.5.1

Submanifold of sphere	Cone is area-minimizing	Cone is stable but not minimizing	Cone is Unstable
$S^{r-1} \times S^{s-1}$	$r=3,\, s=5$ $r=4,\, s=4$ $r+s>8$	$r=2,\, s=6$	$r+s<8$
$\dfrac{SO(2) \times SO(m)}{\mathbf{Z}_2 \times SO(m-2)}$	$m>8$	$m=8$	$3 \le m \le 7$
$\dfrac{SU(2) \times SU(m)}{T^1 \times SU(m-2)}$	$m \ge 4$		$m=2,3$
$\dfrac{Sp(2) \times Sp(m)}{Sp(1)^2 \times Sp(m-2)}$	$m \ge 2$		
$\dfrac{U(5)}{SU(2) \times SU(2) \times T^1}$	×		
$\dfrac{SO(3)}{\mathbf{Z}_2 \times \mathbf{Z}_2}$			×
$\dfrac{SU(3)}{T^2}$			×
$\dfrac{Sp(3)}{Sp(1)^3}$	×		
$\dfrac{Sp(2)}{T^2}$			
$\dfrac{G_2}{T^2}$			
$\dfrac{F_4}{Spin(8)}$	×		
$\dfrac{Spin(10) \times U(1)}{SU(4) \times T^1}$	×		

Chapter 6: Some Perturbation Results

In this chapter we prove some perturbation results. The main result is that certain singular minimal surfaces are area-minimizing in a small neighborhood of the singularity. For codimension 1 surfaces, this result was obtained in [HS]. Another result is an example of a boundary B such that any small perturbation of B bounds an area-minimizing surface with a singularity. In contrast, Hardt and Simon [HS, Theorem 2.1] proved that if C is a *hypercone*, then many small perturbations of the boundary of C_1 give area-minimizing surfaces with no singularities at all.

Motivation

Any area-minimizing surface is minimal. Though the converse is not true, under some conditions we can get a local converse. For example, it is known that if p is a regular point of a minimal surface S, and U is a small enough ball centered at p, then $S \cap U$ is area-minimizing (see [M3, Cor. 3.4] and [F2]). That is, a minimal surface is "locally minimizing" at regular points.

A minimal surface is not always locally minimizing at a singularity. For example, if S is a pair of planes through the origin in $\mathbf{R}^3$, then S is not locally minimizing at the origin. However, if the tangent cone C at p is area-minimizing, then in many cases we can prove that S is locally minimizing at p. In particular, if C has multiplicity one, and if Theorem 1.2.1 holds "strictly" for C, then S is locally minimizing at p.

Overview of proof

The main elements of the proof are as follows:

(1) First, Leon Simon has proved that for S and C as above, S converges to C in the $C^{2,1}$-norm. Thus, if U is a small ball centered at p, then $U \cap S$ will be a $C^{2,1}$-small graph over $U \cap C$.

(2) We modify the methods of Chapter 2 to prove that a $C^{2,1}$-small minimal graph S over the truncated cone $C_1 = C \cap \mathbf{B}(1)$ must be area-minimizing:

(a) We let ϕ be a calibration of C whose comass is strictly less than 1 at points not on C.

87

(b) Using a certain extension of the map from C_1 to S, we slightly deform ϕ so that it is dual to tangent planes of S. We prove that the new closed form ϕ' also has comass 1, so that it calibrates S. (Incidentally, ϕ' will be defined not on all of $\mathbf{R}^n$, but at least on the convex hull of ∂S, which is sufficient.) Since we are working with calibrations rather than retractions, we must assume to begin with that C_1 is orientable. However, we later will make a remark to show that we can allow C_1 and its comparison surface candidates to be unorientable.

Section 6.1: The strict curvature criterion

If a cone passes the curvature criterion (Theorem 1.2.1), and is thus area-minimizing, then it almost certainly passes with some leeway to spare. That is, there is a calibration of the cone which has comass strictly less than 1, at points not on the cone.

6.1.1 Definition: Let C be a k-dimensional area-minimizing cone in $\mathbf{R}^n$ which satisfies the curvature criterion, Theorem 1.2.1. We say that the curvature criterion "holds strictly" if there is some leeway; that is to say,

(1) We can produce a "δ-calibration" of C, a calibration whose comass at a point $q \in \mathbf{R}^n$ is at most $1 - \delta t^2$, where δ is some small number, θ is the angle between the vector q and the cone, and $t = \tan\theta$. This corresponds to replacing the differential inequality in 2.3.8 with

$$(g - \frac{t}{k}g')^2 + (\frac{g'}{k})^2 \leq \left((1 - \delta t^2)\inf_\nu \det(I - t\mathbf{h}_{ij}^\nu)\right)^2.$$

(2) The *closures* of the normal wedges intersect only at 0.

(Note: Condition (2) actually follows from condition (1) above, since the factor $1 - \delta t^2$ widens the normal wedges. However, I didn't feel that the proof of this was worth including, especially since condition (2) can easily be verified directly in practice whenever it holds.)

Any cone which we verify to be area-minimizing using Theorem 1.3.5 and Table 1.4.1 strictly satisfies the curvature criterion; this is inherent in the numerical methods used to compute Table 1.4.1.

Section 6.2: Calibration of a nearby minimal graph

First let us define what we mean by a $C^{2,1}$-small graph over a cone. The main point that requires clarification is what happens near the origin.

6.2.1 Definition: Let C_1 be a truncated cone $C \cap \mathbf{B}^n(1)$; let $\partial C_1 = B \subset \mathbf{S}^{n-1}$. Let f be a function mapping C_1 to a surface $S \subset \mathbf{R}^n$. Then we call S (or f) a "graph over C_1." We let f^v be the associated vector-valued function defined on C_1, given by

$$f^v(p) = f(p) - p.$$

Note: The word "graph" here is actually a misnomer, since we do not require f^v to be normal to C_1.

Our C^2-norm is to be the supremum of the usual C^2-norms over all blowups of the picture:

6.2.2 Definition: Let f be a graph over a truncated cone C_1, with $f(0) = 0$. Let $\|f\|_{2,1/2}$ be the supremum of the usual C^2-norm over points of C lying inside the unit ball, but outside the ball of radius $1/2$. Let

$$f_\alpha(x) = \frac{f(\alpha x)}{\alpha}, \qquad \text{if} \quad 0 < \alpha \leq 1.$$

Let

$$\|f\|_{C^2} = \sup_{\alpha \in (0,1]} \|f_\alpha\|_{2,1/2}.$$

Now we define a $C^{2,1}$-small graph over C_1 as being a $C^{2,1}$ graph which has small C^2-norm. The existence a large family of minimal surfaces which are $C^{2,1}$-small graphs over a truncated cone over a minimal submanifold of the unit sphere is established in [CHS] (see the Introduction, 3.1, and Section 5).

Having defined our small graphs, we turn to the problem of calibrating them. Let C_1 be a truncated cone which strictly satisfies the curvature criterion, and let ϕ be a δ-calibration of C_1 (cf. Definition 6.1.1). Let f be a $C^{2,1}$-small map defined on C_1, whose image is a nearby minimal surface S. We will construct a closed form ω which equals 1 on the tangent planes of S, and which, if $\|f\|_{C^2}$ is small enough, will have comass 1; thus, ω will be a calibration of S.

Extension of the map $f : C_1 \to S$

We construct ω as a pull-back of ϕ. Thus, we need to extend f so that its domain is the support of ϕ. We do this in the same way as we constructed "isonormal coordinates" in Definition 2.3.1.

6.2.3 Definition: Let $f : C_1 \to S$. Let Ω be the support of the calibration ϕ. For each $p \in C_1$, let $\xi(p)$ be the intersection of Ω with the normal space to C_1 at p. Let $\tilde{\xi}$ be slightly larger than ξ, so that the disks $\tilde{\xi}$ still do not intersect (we can do this because the closures of the normal wedges associated with ϕ intersect only at the origin). Let F be an extension of f, defined by

(a) if $p \in C_1$, then $F(p) = f(p)$

(b) F maps the $n - k$ dimensional disk $\tilde{\xi}$ isometrically into the normal space to S at $f(p)$.

This involves a choice of rotation of $\tilde{\xi}$; any $C^{1,1}$ choice will do.

Now if $\|f\|_{C^2}$ is small enough then F will be one-to-one. Thus, we can define

$$G = F^{-1}.$$

The calibration of S will approximately be the pull-back $G^\sharp \phi$. Applying this form to the tangent planes of S, we have

$$G^\sharp \phi(T_q(S)) = (J_q(f))^{-1} \approx 1,$$

where $J_q(f)$ is the Jacobian function $\det(Df)$. However, since $f : C_1 \to S$ need not preserve area, $J_q(f) \neq 1$ in general, so that $G^\sharp \phi$ will not calibrate S. To compensate for the nonunit Jacobian, we adjust ϕ before pulling it back. We do this by multiplying ϕ by a certain real-valued function h. In order to ensure that our new form is still closed, we use the following lemma.

6.2.4 Lemma: Let ϕ be a simple closed differential k-form in $\mathbf{R}^n$. Suppose that $h : \mathbf{R}^n \to \mathbf{R}$ is a function whose directional derivative at p, in the direction X, is 0, whenever X is perpendicular to the dual of ϕ at p. Then the form $h\phi$ is also closed.

Proof: We compute

$$d(h\phi) = dh \wedge \phi + h\, d\phi = dh \wedge \phi.$$

But $dh\big|_p$ is dual to the vector direction in which h increases most rapidly. This direction lies in the k-plane dual to ϕ at p, so that $dh \wedge \phi = 0$. ∎

6.2.5 Definition (Calibration of S): Let $h : C_1 \to \mathbf{R}$ be the Jacobian function

$$h(p) = \det(Df)\big|_p.$$

For each $p \in C_1$, let S_p be the surface of retraction $\Pi^{-1}(p)$ as defined in 1.1.5. Extend h to the support of ϕ by letting it be constant on each S_p. Let

$$\omega = G^{\sharp}(h \cdot \phi).$$

The claim is that ω calibrates S. By Proposition 2.4.1, S_p is completely orthogonal to the dual k-planes of ϕ. Then by Lemma 6.2.4, $h\phi$ is closed, so that ω is closed as well. By Corollary A9 in the Appendix, if $\|\omega\| = 1$, then although it is discontinuous, ω can be used as a calibration. Using Lemma 2.3.3, we see that ω is dual to the tangent planes of S. By construction, $\omega(T_q(S)) = 1$. It remains to show that $\|\omega\| \leq 1$ everywhere.

Estimation of the comass

First note that it suffices to compute the comass at all points $F(p + t\nu)$ for $p \in B$. To see this, let $a < 1$ and $z = a(p + t\nu)$. Then $\omega|_{F(z)}$ is the same as the form we would get at $F(p + t\nu)$ if we replaced f by f_a, where $f_a(p) = \frac{1}{a}f(ap)$ (and thus S would be replaced by a subset of the dilated surface $\frac{1}{a}S$). By our definition of the C^2-norm of f, the same inequalities hold for that form as for $\omega|_{F(z)}$.

The idea of the comass estimate is to first show that the ratio

$$\frac{\|\phi|_{p+t\nu}\|}{\|\phi|_p\|}$$

is only slightly changed by multiplying ϕ by the function h; second, we show that the comass change due to $G^{\sharp}$ is almost the same at $p + t\nu$ as at p. These factors will be small enough so that we will still have $\|\omega\| \leq 1$ at each point, because of the fact that $\|\phi\|$ decreases as we move away from C_1.

6.2.6 Lemma: Let p be in C_1 and ν a unit normal at p, and let $z = p + t\nu$ be a point in the support of the δ-calibration ϕ (cf. Definition 6.1.1). Let h be the (extended) Jacobian function defined as above. Then

$$\left| 1 - \frac{h(p + t\nu))}{h(p)} \right| \leq \frac{\delta}{3} t^2,$$

if ϵ is small compared with δ.

Proof: Recall from Section 2.4 that $g(t) = (r \cos \theta)^{-k}$ on the projection curve γ through p. Thus, the retraction Π of Chapter 1 sends $p + t\nu$ to $(g(t)^{1/k})p$. Since g is C^2 at 0, $g^{1/k} = 1 - O(t^2)$. Then

$$h(p + t\nu) = h\big((g(t)^{1/k})p\big)$$

while

$$|p - (g(t)^{1/k})p| = O(t^2).$$

The norm $\epsilon = \|f\|_{C^2}$ gives a bound on the rate of change of the Jacobian function h as we move around on C_1; if ϵ is small compared with δ then

$$\left| 1 - \frac{h(p + t\nu)}{h(p)} \right| = O(\epsilon t^2) < \frac{\delta}{3} t^2. \quad \blacksquare$$

We now need to see how $G^\sharp$ affects the comass. In the following explanatory paragraph, for clarity let us assume that f is area-preserving, so that $h \equiv 1$.

If DG increased the area of a k-plane (k-vector) near S by at most a factor of $1 + O(\epsilon t^2)$, we would be finished. The difficulty is that the area of some k-planes, depending on their direction, may be multiplied by $1 + O(\epsilon t)$, which is not good enough. In fact, the general situation is that because S is minimal, if a k-plane at $q + t\nu$ is parallel to $T_q(S)$, then DG multiplies its area by at most $1 + O(\epsilon t^2)$, for small t. If one of the directions in $T_q(S)$ is replaced by a vector normal to S at q, then DG may increase the area by $1 + O(\epsilon t)$. To remedy this, we must use the fact that the dual of ω is close to being parallel to S, at points near S.

Stretch factors of the map F

Let $p \in \partial C_1$ and $q = f(p)$. Let ν be normal to C_1 at p, and $\nu' = DF(\nu)$. Let $x = p + t\nu$ and $z = f(x) = q + t\nu'$. Let π and π' be orthogonal projections onto C and S.

To estimate the comass of ω at z, we begin with a lemma telling how much DG stretches a k-plane parallel to S, and one not parallel to S.

6.2.7 Lemma: Let $p \in B$, $q = f(p) \in \partial S$, and $z = q + t\nu' \in \operatorname{spt}(\omega)$. Let $v_1, \ldots, v_k$ be an orthonormal basis for the k-plane at z which is parallel to $T_q(S)$. Let $\xi = v_1 \wedge \cdots v_k$. Let $\epsilon = \|f\|_{C^2}$. Let ξ' be the component of $DG(\xi)$ parallel to $T_p(C)$. There is a constant c independent of q, ϵ, and t, such that

$$\left| h(p)\|\xi'\| - 1 \right| < c\epsilon t^2.$$

Further, if ξ is not parallel to $T_q(S)$, then the same formula holds if we replace $c\epsilon t^2$ by $c\epsilon t$.

Proof: There is some vectorfield $\nu(p)$ normal to C such that if we let $\sigma : p \to p + t\nu(p)$, then

$$DG(\xi) = D\sigma \circ D(f^{-1}) \circ D\pi'(\xi).$$

Using Lemma 2.3.4,

$$D\pi'(\xi) = \det\left(I - t\mathbf{h}'_{ij}{}^{\nu'}\right)^{-1} \cdot T_q(S),$$

and

$$D\sigma\left(T_p(C)\right) = \det\left(I - t\mathbf{h}^\nu_{ij}\right) \cdot T_q(C) + Normal,$$

where $\mathbf{h}$ and $\mathbf{h}'$ are the second fundamental forms of C and S at p and q, respectively. "Normal" denotes terms which include vectors normal to C and to ν at p. (The formula for $D\sigma(T_p(C))$ follows from Lemma 2.3.4 using the fact that $\pi \circ \sigma$ is the identity; we could also derive it from Lemma 2.3.2).

Since $D(f^{-1})$ changes area by a factor of $\frac{1}{h(p)}$, the component ξ' of $DG(\xi)$ parallel to $T_p(C)$ is

$$\frac{\det\left(I - t\mathbf{h}^\nu_{ij}\right)}{h(p)\det\left(I - t\mathbf{h}'_{ij}{}^{\nu'}\right)^{-1}} \, T_p(C).$$

Because C and S are both minimal, these determinants are both $1 - O(t^2)$. Their second-order terms differ by $O(\epsilon)$, because the second fundamental forms $\mathbf{h}$ and $\mathbf{h}'$ differ only by $O(\|f\|_{C^2}) = O(\epsilon)$. Neither determinant reaches zero on the interval of t with which we are concerned (recall that $z = q + t\nu \in \mathrm{spt}(\omega)$), so there is a constant c such that

$$\left| h(p)\|\xi'\| - 1 \right| < c\epsilon t^2.$$

By compactness, we can make c independent of $q \in \partial S$.

To prove the second statement of the lemma, we again use the fact that $\mathbf{h}$ and $\mathbf{h}'$ are nearly the same; by Lemma 2.3.2 and the formula

$$DG(v) = D\sigma \circ D(f^{-1}) \circ D\pi'(v),$$

if v is any unit vector at q, then the component of $DG(v)$ parallel to $T_p(C)$ has length $1 + O(\epsilon t)$. ∎

The k-plane dual to ω

Now we need to know which k-plane is dual to ω at z. For this, we use Lemma 2.3.3:

(1) Let ζ be the $k-1$ plane tangent to B at p. Then $\phi|_z$ is dual to

$$\xi = \zeta \wedge (\alpha(t)\frac{\partial}{\partial r} + \beta(t)\nu,$$

where

$$\alpha(t) = g(t) - \frac{t}{k}g'(t) = 1 - O(t^2)$$

and

$$\beta(t) = \frac{1}{k}g'(t) = O(t).$$

(2) The orthogonal complement of ξ is

$$N \wedge (\beta\frac{\partial}{\partial r} - \alpha\nu),$$

where N is the wedge product of all normals to C at p except ν.

(3) DF maps $\xi^\perp$ to

$$N' \wedge (\beta(t)DF(\frac{\partial}{\partial r}) - \alpha\nu')$$

where N' is the wedge product of all normals to S at q except ν'.

Since F preserves distance from C and S, it is clear that $DF(\frac{\partial}{\partial r})$ has no component in the direction ν'. When wedged with N', the other normal components go away, so DF maps $\xi^{\perp}$ to

$$N' \wedge (\beta u - \alpha \nu')$$

for some nearly-unit vector u tangent to S at q.

(4) The orthogonal complement of $N' \wedge (\beta u - \alpha \nu')$ is

$$\xi' = \zeta' \wedge (\alpha u + \beta \nu')$$

where ζ' is a unit simple $k-1$ vector (or $k-1$ plane) contained in the k-plane parallel to S at q, and $u \perp \zeta'$; thus, $\zeta' \wedge u$ is parallel to S at q.

Therefore, ξ' is the dual of ω at z.

It is convenient to normalize these vectors; let

$$\xi'' = \zeta' \wedge (\alpha' u' + \beta' \nu')$$

where $(\alpha' u' + \beta' \nu')$ is a unit length multiple of $(\alpha u + \beta \nu')$, and u' has unit length. Then

$$\alpha'^2 + \beta'^2 = 1,$$

and ξ'' is a unit k-vector dual to ω at z. Since u has almost-unit length, we have not changed α and β by very much; it is still true that

$$\alpha' = 1 - O(t^2) \quad \text{and} \quad \beta' = O(t).$$

Applying ω to its unit dual k-plane ξ''

Now let

$$\xi_1 = \zeta' \wedge u'$$

and

$$\xi_2 = \zeta' \wedge \nu'.$$

Then

$$\omega(\xi'') = \alpha' \omega(\xi_1) + \beta' \omega(\xi_2)$$

$$= h\phi\Big(DG(\zeta') \wedge (\alpha'DG(u') + \beta'DG(\nu'))\Big).$$

Let P be the part of $DG(\zeta')$ parallel to $T_p(C)$, and Y the part of $DG(u')$ parallel to $T_p(C)$. Then

$$\omega(\xi'') = h\phi(P \wedge (\alpha'Y + \beta'\nu)).$$

Since $Y \perp \nu$ and $P \perp \nu$, we have

$$\omega(\xi'') \le h(G(z))(1 - \delta t^2)\sqrt{\alpha'^2\|P \wedge Y\|^2 + \beta'^2\|P \wedge \nu\|^2}$$

$$\le h(p)(1 - \frac{\delta}{2}t^2)\sqrt{\alpha'^2\|P \wedge Y\|^2 + \beta'^2\|P \wedge \nu\|^2}$$

by Lemma 6.2.6;

$$= (1 - \frac{\delta}{2}t^2)\sqrt{1 - \alpha'^2(1 - h(p)^2\|P \wedge Y\|^2) - \beta'^2(1 - h(p)^2\|P \wedge \nu\|^2)},$$

using the fact that $\alpha'^2 + \beta'^2 = 1$.

Let c_1 be a constant such that $\beta'(t) < c_1 t$ for all $t > 0$. Let $\epsilon = \|f\|_{C^2}$. By Lemma 6.2.7, there is a constant c_2 such that

$$\left|1 - h(p)^2\|P \wedge Y\|^2\right| < c_2\epsilon t^2$$

and

$$\left|1 - h(p)^2\|P \wedge \nu\|^2\right| < c_2\epsilon t.$$

Then

$$\omega(\xi'') \le (1 - \frac{\delta}{2}t^2)\sqrt{1 + \alpha'^2 c_2\epsilon t^2 + \beta'^2 c_2\epsilon t}$$

$$\le (1 - \frac{\delta}{2}t^2)\sqrt{1 + c_2\epsilon(t^2 + c_1 t^3)} \le 1$$

if ϵ is small compared to δ.

The domain of ω

Now since the calibration ω does not exist in all of $\mathbf{R}^n$, we need to make sure that it exists on the convex hull H of ∂S. This is sufficient; see Lemma A4 in the appendix.

We ensure that ω exists on all of H by letting it be zero everywhere on the set

$$M = H \backslash f(\text{spt}(\phi));$$

that is, if z is a point in H, but ω is not yet defined at z, then we let $\omega|_z = 0$. The form ω will still qualify as a calibration as long as the points $q + t\nu$, for $q \in \partial S$ and ν normal to S at q, do not lie on the boundary of M; this will mean that $\omega = 0$ at all points of ∂M already, so that we will not introduce any new singularities in ω

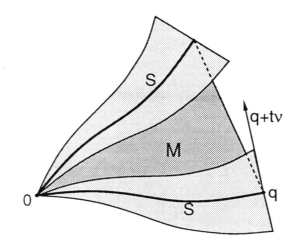

by setting it equal to zero on M.

Let $q \in \partial S$. By Lemma A5 in the appendix, each normal line $q + t\nu$ lies in a hyperplane ξ, such that all points of ∂S (except q) lie strictly on one side of ξ. This tells us that H lies on one side of ξ, and only intersects the line $q + t\nu$ at q. This tells us that the only possible intersection of ∂M with the line $q + t\nu$ is at q. To show that $\partial S \cap \partial M$ is empty, suppose that z is a point near ∂S, with $z \in H$. Let q be the nearest point to z in S. By Corollary A7 in the appendix, $q \notin \partial S$. But this means that ω is already defined and nonzero at z, since $z - q$ is a small vector normal to S at q.

We now have the following theorem.

6.2.8 Theorem: Let C be a cone for which the curvature criterion holds strictly (cf. Definition 6.1.1). Let $S = f(C_1)$ be a minimal graph over the truncated cone C_1, and $\epsilon = \|f\|_{C^2}$ (cf. Definitions 6.2.1 and 6.2.2). If ϵ is small enough, then S is area-minimizing.

Proof: If C is orientable, we can calibrate S as described above, with the form ω. If C is unorientable, or if we wish to allow unorientable comparison surfaces, we use a retraction argument. Let

$$\Pi' = F \circ \Pi \circ G,$$

where Π is the retraction orthogonal to ϕ (which still exists locally), constructed as in Chapter 1 (see Definition 1.1.5). By Lemma 2.3.3, the surfaces of retraction of Π' are orthogonal to the dual k-planes of the calibration ω (which are the same as the duals of ω). By Lemma 2.4.2, since the comass of the locally-defined form ω is 1 on $T_q(S)$ and less than 1 elsewhere, Π' is an area-decreasing retraction. ∎

Section 6.3: Minimal surfaces in a neighborhood of a singularity

Now suppose that S is a minimal surface with a singularity at 0. Let C be the tangent cone to S at 0; suppose that C is multiplicity 1 and has an isolated singularity at 0. By Theorem 5 of [S], inside a neighborhood of 0 we can write S as a $C^{2,1}$ graph over C. Further, by making the neighborhood small, we can make the C^2-norm of the graph small. More precisely, if $\epsilon > 0$ then there is a dilation S_α of S and a map f such that $\|f\|_{C^2} < \epsilon$, and

$$S_\alpha \cap \mathbf{B}^n(1) = f\big(C \cap \mathbf{B}^n(1)\big).$$

By Theorem 6.2.8, if we have chosen ϵ small enough then we can calibrate S_α. Therefore, we have the following theorem.

6.3.1 Theorem: Let S be a minimal surface with a singularity at $p \in \mathbf{R}^n$. Let C be the tangent cone to S at p. If C has multiplicity 1 and strictly satisfies the curvature criterion Theorem 1.2.1, then in some neighborhood of p, S is area-minimizing. ∎

Section 6.4: A persistent singularity in a minimizing surface

Now we give an example of a surface with a singularity at 0, such that for any small perturbation of the boundary, the area-minimizing surface with the new boundary still has a singularity near 0.

6.4.1 Theorem: Let $C = C_1 + C_2$ be the union of two completely orthogonal unit 3-balls in $\mathbf{R}^6$. Let $B_i = \partial C_i$, and $B = B_1 + B_2$. Let $\Sigma = \Sigma_1 + \Sigma_2$ be a $C^{3,\alpha}$ perturbation of B, and let S be an area-minimizing surface with boundary Σ. Then if the $C^{3,\alpha}$-norm of the perturbation is sufficiently small, S is unique, and has a singularity near 0.

Proof: By the implicit function theorem of partial differential equations, working with the minimal surface system of equations and using the $C^{3,\alpha}$ norm, there is a minimal surface S_1 with boundary Σ_1, which is a $C^{3,\alpha}$-small graph $f_1(C_1)$. Likewise, let S_2 be a $C^{3,\alpha}$-small minimal graph $f_2(C_2)$, with boundary Σ_2. Then $S_1 \cap S_2 = \{z\}$, a single point near 0. By a $C^{3,\alpha}$-small shift in the maps f_1 and f_2, we can make $f_1(0) = f_2(0) = z$; then we can define a new map f as equaling f_1 on C_1 and f_2 on C_2. If we subtract the vector z from each point of S, the resulting surface is still a $C^{3,\alpha}$-small graph over C. Because of this, we can assume that $f(0) = 0$.

Now C strictly satisfies the curvature criterion. Since f is a $C^{3,\alpha}$-small (and thus $C^{2,1}$-small) graph over C and S is minimal, S is calibrated by a form which is 1 *only* on the tangent planes of S. Thus, S is the unique area-minimizing surface with boundary Σ. ∎

Chapter 7: Open Questions

(1) Are there connected, isoparametric cones of arbitrarily large dimension which are minimal but not area-minimizing? The highest dimensional example I know of is the 15-dimensional cone in $\mathbf{R}^{16}$ which is noted in Theorem 5.5.2. A closely-related question is whether the curvature of an isoparametric cone in a given dimension can be arbitrarily large.

(2) A k-covector is called "simple" if it is dual to a k-vector which can be represented as a k-plane (i.e., a k-vector which is a wedge product of vectors.) If we view the method of this paper in terms of calibrations (as in Chapter 2), the calibrations are simple, with nonconstant coefficients, and are constructed in a local manner. Other important calibrations are nonsimple, constant-coefficient calibrations (see [HL], for example). In his thesis, Benny Cheng constructs nonsimple, nonconstant-coefficient calibrations. Can the method of this paper be improved by using nonconstant-coefficient calibrations, constructed in the same local fashion as in Chapter 2, but using more information from curvature, or perhaps the holonomy of the normal bundle? It is useful to examine the special Lagrangian form (see [HL]), which calibrates certain area-minimizing cones for which the method of this paper fails — for example, certain pairs of k-planes, as well as a 3-dimensional cone in $\mathbf{R}^6$ described in [HL], invariant under the action of the maximal torus in $\mathbf{SU}(n)$.

(3) Which connected area-minimizing isoparametric cones are *not* strictly minimizing (see [HS] for definition and discussion; a k-dimensional truncated cone whose area is A is strictly minimizing if there is an α such that for each $\epsilon > 0$, any surface with the same boundary as the truncated cone which stays outside the ball of radius ϵ centered at $\mathbf{0}$, must have area greater than $A(1 + \alpha\,\epsilon^k)$.) The only ones I know are a line, a 2-plane, and Harvey and Lawson's 3-dimensional cone mentioned in (2) above. Other possibilities might be cones calibrated by constant-coefficient calibrations such as the special Lagrangian form and the associative and coassociative forms (see [ChB, historical notes] and [HM]). Whenever the criterion of this paper holds, the cone is strictly minimizing.

100

(4) Is it true that if a cone is strictly area-minimizing, then any minimal surface with that (unique) tangent cone is area-minimizing inside some neighborhood of the singularity? (Generalizing the results of Chapter 6).

(5) Besides minimizing area, we can also ask which surfaces minimize other integrands. Such questions arise in crystal formations, for example. What singularities can occur? Frank Morgan [M9] has recently proved that the non-area-minimizing (in fact, unstable) cone over $S^1 \times S^1$ is minimizing with respect to a different elliptic integrand. Can the method of this paper be modified to shed light on these other integrands?

(6) This paper deals only with isolated singularities. If we cross a cone with a line, it becomes evident how to generalize the proof to show that such a cone is minimizing (if we were able to do the original cone by our method). Can we generalize to more complicated singularities?

(7) Some other questions are raised in Chapter 5 of this paper; see Section 5.3 and the end of Section 5.2.

Appendix

The following lemma and corollaries are technical results used in simplifying the hypotheses of Theorem 1.2.1, in order to get Theorem 1.3.5, which is more readily applicable.

A1 Lemma: Let $m \geq 2$, $0 \leq t \leq \sqrt{\frac{m}{m-1}}$, and $a_i \in \mathbf{R}$. Subject to the restrictions $\sum_{i=1}^{m} a_i = 0$ and $\sum_{i=1}^{m} a_i^2 = 1$, the minimum

$$\min_{\{a_i\}} \prod_{i=1}^{m} (1 - a_i t)$$

is given by letting $a_1 = \sqrt{\frac{m-1}{m}}$ and $a_i = -\sqrt{\frac{1}{m(m-1)}}$ for $i > 1$.

Proof: First we will see that no three of the a_i can be different: Suppose the set $\{a_i\}$ gives the desired minimum. Then in particular, $(1 - a_1 t)(1 - a_2 t)(1 - a_3 t)$ is the smallest it can be under the restrictions

$$a_1 + a_2 + a_3 = c_1$$

$$a_1^2 + a_2^2 + a_3^2 = c_2$$

for some pair of constants $c1$, $c2$. Thus,

$$(1 - a_1 t)(1 - a_2 t)(1 - a_3 t) = 1 - c_1 t + \frac{1}{2}(c_1^2 - c_2)t^2 - a_1 a_2 a_3 t^3,$$

which is minimized for all $t \geq 0$ by maximizing $a_1 a_2 a_3$. By the method of Lagrange, we require the matrix

$$\begin{pmatrix} 1 & 1 & 1 \\ 2a_1 & 2a_2 & 2a_3 \\ a_2 a_3 & a_1 a_3 & a_1 a_2 \end{pmatrix}$$

to be singular, i.e.,

$$2(a_2^2 a_1 + a_3^2 a_2 + a_1^2 a_3 - a_3^2 a_1 - a_2^2 a_3 - a_1^2 a_2) = 2(a_1 - a_2)(a_2 - a_3)(a_3 - a_1) = 0.$$

Then a_1, a_2, and a_3 cannot all be different.

The same holds true for any triple a_i, a_j, a_k, so there are only two different values represented among all of the a_i. Together with the restrictions $\sum a_i = 0$

102

and $\sum a_i^2 = 1$ this will completely determine the solution, once we know how many should be positive and how many negative. Suppose

$$a_1 = \cdots = a_r = c > 0$$

and

$$a_{r+1} = \cdots = a_m = -d < 0.$$

Then

$$\sum a_i = 0 \Rightarrow cr = ds,$$

$$1 = \sum a_i^2 = rc^2 + (m-r)d^2 \Rightarrow d = \sqrt{\frac{r}{m(m-r)}}, c = \sqrt{\frac{m-r}{mr}}$$

$$\prod(1 - a_i t) = \left(1 - t\sqrt{\frac{m-r}{mr}}\right)^r \left(1 + t\sqrt{\frac{r}{m(m-r)}}\right)^{m-r},$$

which we wish to minimize over all integers $r \in [1, m-1]$. The correct choice is $r = 1$, which we prove below by changing r to a continuous variable and differentiating:

With $m \geq 3$ and $t \in [0, \sqrt{\frac{m}{m-1}}]$ as above, let

$$f(x) = \left(1 - t\sqrt{\frac{m-x}{mx}}\right)^x \left(1 + t\sqrt{\frac{x}{m(m-x)}}\right)^{m-x}.$$

Then for $x \in [1, m-1]$, $\frac{df}{dx} \geq 0$.

To prove this, let $g(x) = \ln(f(x))$; it suffices to show that $\frac{dg}{dx} \geq 0$.

$$g(x) = x \ln\left(1 - t\sqrt{\frac{m-x}{mx}}\right) + (m-x)\ln\left(1 + t\sqrt{\frac{x}{m(m-x)}}\right)$$

$$g'(x) = \frac{-\frac{1}{2}xt\sqrt{\frac{mx}{m-x}}\left(\frac{-mx-m(m-x)}{m^2x^2}\right)}{1 - t\sqrt{\frac{m-x}{mx}}} + \ln\left(1 - t\sqrt{\frac{m-x}{mx}}\right)$$

$$+ \frac{\frac{1}{2}t(m-x)\sqrt{\frac{m(m-x)}{x}}\left(\frac{m^2-mx+mx}{m^2(m-x)^2}\right)}{1 + t\sqrt{\frac{x}{m(m-x)}}} - \ln\left(1 + t\sqrt{\frac{x}{m(m-x)}}\right)$$

$$= \frac{\frac{1}{2}t\sqrt{\frac{m}{x(m-x)}}}{1 - t\sqrt{\frac{m-x}{mx}}} + \frac{\frac{1}{2}t\sqrt{\frac{m}{x(m-x)}}}{1 + t\sqrt{\frac{x}{m(m-x)}}} + \ln\left(1 - \frac{mt}{\sqrt{m(m-x)x} + xt}\right).$$

Let $a = mt$, and $b = \sqrt{m(m-x)x} + xt$. Then

$$2g'(x) = 2\ln(1 - \frac{a}{b}) + \frac{a}{b} + \frac{a}{(b-a)}.$$

Since $0 < t < \sqrt{\frac{mx}{(m-x)}}$,

$$(m-x)t < \sqrt{m(m-x)x},$$

so $0 < a < b$.

To show

$$2\ln(1 - \frac{a}{b}) + \frac{a}{b} + \frac{a}{(b-a)} \geq 0,$$

note that it is true for $a = 0$, and take

$$\frac{\partial}{\partial a} = \frac{-2}{b-a} + \frac{1}{b} + \frac{b}{(b-a)^2}$$

$$= \frac{1}{b(b-a)^2}(-2b(b-a) + b^2 - 2ab + a^2 + b^2)$$

$$= \frac{a^2}{b(b-a)^2} \geq 0. \quad \blacksquare$$

A2 Corollary: Let $\alpha > 0$, $m \geq 3$, $0 \leq t \leq \frac{1}{\alpha}\sqrt{\frac{m}{m-1}}$, and $a_i \in \mathbf{R}$. Subject to the restrictions $\sum_{i=1}^{m} a_i = 0$ and $\sum_{i=1}^{m} a_i^2 = \alpha^2$, the minimum

$$\min_{\{a_i\}} \prod_{i=1}^{m}(1 - a_i t)$$

is given by letting $a_1 = \alpha\sqrt{\frac{m-1}{m}}$ and $a_i = -\alpha\sqrt{\frac{1}{m(m-1)}}$ for $i > 1$. $\quad \blacksquare$

A3 Corollary: For any fixed $t \in [0,1]$, the quantity

$$g(m,t) = \left(1 - t\sqrt{\frac{m-1}{m}}\right)\left(1 + \frac{t}{\sqrt{m(m-1)}}\right)^{m-1}$$

is a nonincreasing function of the integer variable m.

Proof: The proof is by induction on m. By Lemma A1,

$$g(m,t) = \inf_{\{a_i\}} \prod_{i=1}^{m}(1 - a_i t),$$

subject to

$$\sum_{i=1}^{m} a_i = 0 \quad \text{and} \quad \sum_{i=1}^{m} a_i^2 = 1.$$

By adding one more variable a_{m+1}, we can only decrease this minimum, since the choice

$$a_1 = \sqrt{\frac{m-1}{m}}, \quad a_2 = \cdots = a_m = -\sqrt{\frac{1}{m(m-1)}}, \quad \text{and} \quad a_{m+1} = 0$$

satisfies the restrictions and gives

$$\prod_{i=1}^{m+1} (1 - a_i t) = g(m, t). \quad \blacksquare$$

The following results are used in proving that the domain of a certain calibration in Chapter 6 is large enough.

A4 Lemma (domain of a calibration): Let S be a compact surface with boundary in $\mathbf{R}^n$. Let H be the convex hull of ∂S. Suppose that ω is a calibration of S, and that ω is defined on all of H. Then this domain for ω is large enough to ensure that S is area-minimizing.

Proof: Let T be an area-minimizing surface with boundary ∂S. Let $z \in H$, and $U = z \text{✕} (T - S)$. Then $\partial U = T - S$ and $U \in H$, so that

$$\int_{T-S} \omega = \int_U d\omega = 0$$

$$\text{Area}(T) \geq \int_T \omega = \int_S \omega = \text{Area}(S)$$

since ω calibrates S. Thus, S is area-minimizing.

A5 Lemma: Let C_1 be a truncated cone $C \cap \mathbf{B}^n(1)$; let $\partial C = B \subset \mathbf{S}^{n-1}$. Let $S = f(C_1)$ be a C^2-small minimal graph over C_1, with $\Sigma = f(B)$. We do not require $\Sigma \subset \mathbf{S}^{n-1}$. Let $p \in B$ and $f(p) = q \in \Sigma$. Let X be the inward tangent to S at q; that is, X is tangent to S and normal to Σ. Let ξ be the hyperplane of $\mathbf{R}^n$ which is normal to X. Then Σ lies on one side of ξ, and only touches ξ at q.

Proof: Let $\epsilon = \|f\|_{C^2}$. Then $f(B)$ is contained in the ball $\mathbf{B}^n(1+\epsilon)$. Let A be the small part of $\mathbf{B}^n(1+\epsilon)$ which lies on one side of ξ. It suffices to prove that $f(B) \cap A$ lies on the same side of ξ as the origin. For this it will be enough to find a geodesic neighborhood $\mathcal{N}$ of p such that $f(B \cap \mathcal{N})$ contains $f(B) \cap A$, and such that for any geodesic γ through p, $(f \circ \gamma)''(s)$ points to the "origin side" of ξ, whenever $\gamma(s) \in \mathcal{N}$. (By "geodesic neighborhood" we mean that for some s_0, the union of geodesics of length s_0 starting at p covers $\mathcal{N}$, and none of them leaves and reenters $\mathcal{N}$).

Now let v be the inward-pointing vector $\overrightarrow{p0}$. Since $B \subset \mathbf{S}^{n-1}$, $\langle \gamma''(s), v \rangle > 0$ for any geodesic γ with $\gamma(0) = p$. Since the space of (unit speed) geodesics through p is compact, there is a positive number ϵ_1 such that $\langle \gamma''(s), v \rangle > \epsilon_1$ for any unit speed geodesic through p.

Again by a compactness argument, we can find a number α such that the $\mathbf{R}^n$-neighborhood $\mathcal{N}$ of p with radius α has the following property:

If γ is a unit speed geodesic through p and $\gamma(s) \in \mathcal{N}$, then $\langle \gamma''(s), v \rangle > \frac{\epsilon_1}{2}$. If $\epsilon = \|f\|_{C^2}$ is small enough, then X will be sufficiently near v that

$$\langle \gamma''(s), X \rangle > \frac{\epsilon_1}{3}.$$

Further, if ϵ is small enough then

$$\langle (f \circ \gamma)''(s), X \rangle > 0$$

whenever $\gamma(s) \in \mathcal{N}$. Finally, if ϵ is small enough, then $f(B) \cap A \subset f(B \cap \mathcal{N})$. ∎

A6 Corollary: The convex hull H of ∂S lies on one side of ξ, and touches ξ only at q. Since $S \subset H$, S also lies on one side of ξ. ∎

A7 Corollary: Let $z \in \mathbf{R}^n$ be a point not on S, which is near enough to S so that the nearest point $p \in S$ is well-defined. If $p \in \partial S$, then z is not in the convex hull H of ∂S.

Proof: Let X be the inward tangent to S at p. Let Y be the vector $z - p$. If $\langle X, Y \rangle$ were positive, then p would not be the nearest point to z. Thus, $\langle X, Y \rangle \leq 0$. This says that z lies on or to the far side of ξ, so that it cannot be in H. ∎

A8 Theorem: The exterior derivative of a Lipschitz form can be used as a calibration. That is, let ω be a Lipschitz $k-1$ form defined on $\mathbf{R}^n$; let $\operatorname{sing}(\omega)$ denote the set on which ω is not C^1. Let $\phi = d\omega$; then ϕ exists $\mathcal{H}^n$-almost everywhere. Suppose that $\|\phi\| \leq 1$ at each point where ϕ is defined. Suppose that there is a k-dimensional surface S such that

$$\mathcal{H}^k(S \cap \operatorname{sing}(\omega)) = 0$$

and

$$\phi(T_p(S)) = 1$$

wherever it is defined. Then S is area-minimizing.

Proof: Our goal is to find a smoothed form $\tilde{\phi} = d(\tilde{\omega})$ whose comass is still at most 1, and which is arbitrarily near 1 on the tangent planes of S. Then if T is a surface such that $\partial T = \partial S$, we will have

$$\text{Area}(S) \leq (1+\epsilon) \int_S \tilde{\phi} = (1+\epsilon) \int_{\partial S} \tilde{\omega} = (1+\epsilon) \int_{\partial T} \tilde{\omega} = (1+\epsilon) \int_T \tilde{\phi} \leq (1+\epsilon)\text{Area}(T).$$

But ϵ will be independent of T, so that the above inequalities force

$$\text{Area}(S) \leq \text{Area}(T).$$

Actually, $\tilde{\phi}$ will be arbitrarily near 1 on nearly all of S rather than on all of S.

The method we will choose to smooth ω and ϕ is called mollification, or regularization (see [GT]).

Definition: Let $\sigma : \mathbf{R}^n \to \mathbf{R}$ be a C^∞ function whose support is contained in the unit ball centered at 0, and whose integral is 1. For $\epsilon \in (0,1)$ let

$$\sigma_\epsilon(z) = \frac{1}{\epsilon^n}\sigma(\epsilon z)$$

so that

$$\int \sigma_\epsilon = 1 \quad \text{and} \quad \operatorname{spt}(\sigma_\epsilon) \in \mathbf{B}^n(\epsilon).$$

Then σ_ϵ is called a mollifier.

Definition: If f is a real-valued function defined almost everywhere in $\mathbf{R}^n$, we define

$$f_\epsilon(x) = \int_{\mathbf{R}^n} f(y)\sigma_\epsilon(y - x).$$

Some properties of mollification, which we will state but not prove, are the following:

(1) If f is a locally bounded, measurable function then f_ϵ is C^∞.

(2) Mollification is linear;

$$(f + g)_\epsilon = f_\epsilon + g_\epsilon,$$

and

$$(af)_\epsilon = a(f_\epsilon).$$

(3) Mollification commutes with differentiation of a Lipschitz function:

$$\frac{\partial}{\partial x^i}(f_\epsilon) = (\frac{\partial}{\partial x^i}f)_\epsilon$$

(see [GT, Lemma 7.3]).

(4) If $c_1 < f < c_2$ on the ϵ-neighborhood of a set K, then $c_1 < f < c_2$ on K.

Now let ω be a Lipschitz $k - 1$ form as stated above, and write

$$\omega = f_I \mathbf{e}^I,$$

where I is a multi-index ranging over all sets of integers $(i_1, \ldots, i_{k-1})$ with $1 \le i_1 < \cdots < i_{k-1} \le n$, and $\mathbf{e}^I$ is the k-covector $\mathbf{e}^{i_1} \wedge \cdots \wedge \mathbf{e}^{i_k-1}$. Similarly, let ϕ be the k-form $d\omega$, written as

$$\phi = g_J \mathbf{e}^J.$$

Since mollification commutes with partial derivatives of Lipschitz functions and mollification is linear, we have

$$d(\omega_\epsilon) = (d\omega)_\epsilon = \phi_\epsilon.$$

The important point here is that ϕ_ϵ is closed.

Next, note that if ξ is a constant k-covectorfield in $\mathbf{R}^n$, and we define

$$h(z) = \phi\big|_z(\xi),$$

then

$$h_\epsilon(z) = \phi_\epsilon\big|_z(\xi).$$

That is, mollification commutes with application of ϕ to a constant k-covector. To see this, write $\xi = \tau^J \mathbf{e}_J$; then

$$\phi(\xi) = \tau^J g_J.$$

Since each function τ^J is constant and mollification is linear,

$$h_\epsilon = (\phi(\xi))_\epsilon = (\tau^J g_J)_\epsilon = \tau^J (g_J)_\epsilon = \phi_\epsilon(\xi).$$

Now we can prove that mollification does not increase comass globally. Recall that $\|\phi\| \le 1$ at each point, and that ϕ is defined almost everywhere. We want to show that $\phi_\epsilon(\xi) \le 1$, for any k-plane ξ at any $z \in \mathbf{R}^n$. Let ξ also represent a constant k-covectorfield; then

$$\phi_\epsilon(\xi) = (\phi(\xi))_\epsilon.$$

But $\phi(\xi) \le 1$ everywhere, so $(\phi(\xi))_\epsilon \le 1$.

Finally, we need to show that ϕ_ϵ is near 1 on most of S, so that

$$\text{Area}(S) \approx \int_S \phi_\epsilon.$$

Since $\text{sing}(\omega) \cap S$ is a closed set having zero $\mathcal{H}^k$ measure, for any ϵ_1 we can find a compact subset $S_c \subset S$ such that ω is C^1 on S_c and

$$\mathcal{H}^k(S) - \mathcal{H}^k(S_c) < \epsilon_1.$$

Choose $\epsilon_2 > 0$ and let δ be small enough so that for each $p \in S_c$ and each neighborhood $\mathcal{N}(p, \delta)$ of radius δ about p,

$$\sup_{\mathcal{N}(p,\delta)} \big|g_J(z) - g_J(p)\big| < \epsilon_2.$$

Then

$$\phi_\delta(T_p(S)) = (g_J)_\delta \langle \mathbf{e}^J, T_p(S) \rangle = 1 + O(\epsilon_2),$$

so that

$$\int_S \phi_\delta = \int_{S_c} \phi_\delta - \int_{S-S_c} \phi_\delta = (1 + O(\epsilon_1))(1 + O(\epsilon_2))\text{Area}(S) - O(\epsilon_1),$$

which we can make as close to Area(S) as we please. Then for any $\epsilon > 0$ we can find δ such that

$$\text{Area}(S) \le (1 + \epsilon) \int_S \phi_\delta = (1 + \epsilon) \int_T \phi_\delta \le (1 + \epsilon)\text{Area}(T). \quad \blacksquare$$

A9 Corollary: If $\phi = d\omega$ where ω is a Lipschitz form, and h is a real-valued $C^{1,1}$ function such that $h\phi$ is still closed wherever it is defined, then $h\phi$ is a candidate for a calibration (i.e., it is a calibration if it has comass 1).

Further, if $G : \mathbf{R}^n \to \mathbf{R}^n$ is a $C^{1,1}$ map then $G^\sharp(h\phi)$ is also a candidate for a calibration.

Proof: We show that $h\phi = d(\psi)$ for some Lipschitz form ψ; then we can use Theorem A8.

Let

$$\psi = h\omega + \eta;$$

then we want to find η such that

$$h\phi = d\psi = d(h\omega + \eta) = h\phi + dh \wedge \omega + d\eta$$

$$-dh \wedge \omega = d\eta.$$

We have reduced the problem to proving that a Lipschitz form which is closed almost everywhere is the derivative of a Lipschitz form η. This is true; it follows from the proof of the Poincaré Lemma (see [Sp, p. 94]).

Now since $h\phi = d\psi$,

$$G^\sharp(h\phi) = G^\sharp(d\psi) = d(G^\sharp\psi),$$

and $G^\sharp\psi$ is Lipschitz. $\quad \blacksquare$

REFERENCES

[A]. Almgren, F.: **Q** Valued Functions Minimizing Dirichlet's Integral and the Regularity of Area Minimizing Currents Up to Codimension 2. See Bulletin AMS **(ns) 8, no. 2** (1983), 327-328

[B]. Bindschadler, D.: Absolutely Area-minimizing Singular Cones of Arbitrary Codimension. Trans. AMS **243** (1978), 223-233

[BDG]. Bombieri, E., DeGiorgi, E., Giusti, E.: Minimal Cones and the Bernstein Problem. Invent. Math. **7** (1969), 243-268

[BD]. Boyce, W., DiPrima, R.: Elementary Differential Equations. New York: John Wiley & Sons 1977

[CHS]. Caffarelli, L., Hardt, R., Simon, L.: Minimal Surfaces with Isolated Singularities. Manuscripta Math. **48 (no. 1-3)** (1984), 1-18

[ChS]. Chang, S.: Two-dimensional Area-Minimizing Integral Currents are Classical Minimal Surfaces. Ph.D. Thesis, Princeton University (1986)

[ChB]. Cheng, B.: Area-Minimizing Equivariant Cones and Coflat Calibrations. Ph.D. Thesis, Mass. Inst. of Technology (1987)

[F1]. Federer, H.: Geometric Measure Theory. New York: Springer-Verlag 1969

[F2]. Federer, H.: A Minimizing Property of Extremal Submanifolds. Archive for Rational Mechanics and Analysis **59** (1975), 207-217

[F3]. Federer, H.: Real Flat Chains, Cochains, and Variational Problems. Indiana Univ. Math. J. **24** (1974/75), 351-407

[Fo]. Fomenko, A. T.: Multidimensional Variational Methods in the Topology of Extremals. Russ. Math. Surveys **36:6** (1981), 127-165

[GT]. Gilbarg, D., Trudinger, N.S.: Elliptic Partial Differential Equations of Second Order, 2nd ed.. New York: Springer-Verlag 1983

[HS]. Hardt, R., Simon, L.: Area Minimizing Hypersurfaces with Isolated Singularities. J. Reine. Angew. Math. **362** (1985), 102-129

[HM]. Harvey, R., Morgan, F.: The Faces of the Grassmannian of 3-planes in $\mathbf{R}^7$ (Calibrated Geometries on $\mathbf{R}^7$). Invent. Math. **83** (1986), 191-228

[HL]. Harvey, R., Lawson, Jr., H. B.: Calibrated Geometries. Acta Mathematica **148** (1982), 47-157

[Hs]. Hsiang, W. Y.: Minimal Cones and the Spherical Bernstein Problem, II.. Invent. Math. **74 (no. 3)** (1983), 351-369

[L1]. Lawlor, G.: The Angle Criterion. Invent. Math. **95** (1989), 437–446

[L2]. Lawlor, G.: Area-minimizing m-tuples of k-planes. To appear.

111

MEMOIRS of the American Mathematical Society

SUBMISSION. This journal is designed particularly for long research papers (and groups of cognate papers) in pure and applied mathematics. The papers, in general, are longer than those in the TRANSACTIONS of the American Mathematical Society, with which it shares an editorial committee. Mathematical papers intended for publication in the Memoirs should be addressed to one of the editors:

Ordinary differential equations, partial differential equations and applied mathematics to ROGER D. NUSSBAUM, Department of Mathematics, Rutgers University, New Brunswick, NJ 08903

Harmonic analysis, representation theory and Lie theory to AVNER D. ASH, Department of Mathematics, The Ohio State University, 231 West 18th Avenue, Columbus, OH 43210

Abstract analysis to MASAMICHI TAKESAKI, Department of Mathematics, University of California, Los Angeles, CA 90024

Real and harmonic analysis to DAVID JERISON, Department of Mathematics, M.I.T., Rm 2–180, Cambridge, MA 02139

Algebra and algebraic geometry to JUDITH D. SALLY, Department of Mathematics, Northwestern University, Evanston, IL 60208

Geometric topology and general topology to JAMES W. CANNON, Department of Mathematics, Brigham Young University, Provo, UT 84602

Algebraic topology and differential topology to RALPH COHEN, Department of Mathematics, Stanford University, Stanford, CA 94305

Global analysis and differential geometry to JERRY L. KAZDAN, Department of Mathematics, University of Pennsylvania, E1, Philadelphia, PA 19104-6395

Probability and statistics to RICHARD DURRETT, Department of Mathematics, Cornell University, Ithaca, NY 14853-7901

Combinatorics and number theory to CARL POMERANCE, Department of Mathematics, University of Georgia, Athens, GA 30602

Logic, set theory, general topology and universal algebra to JAMES E. BAUMGARTNER, Department of Mathematics, Dartmouth College, Hanover, NH 03755

Algebraic number theory, analytic number theory and modular forms to AUDREY TERRAS, Department of Mathematics, University of California at San Diego, La Jolla, CA 92093

Complex analysis and nonlinear partial differential equations to SUN-YUNG A. CHANG, Department of Mathematics, University of California at Los Angeles, Los Angeles, CA 90024

All other communications to the editors should be addressed to the Managing Editor, DAVID J. SALTMAN, Department of Mathematics, University of Texas at Austin, Austin, TX 78713.

General instructions to authors for

PREPARING REPRODUCTION COPY FOR MEMOIRS

> **For more detailed instructions send for AMS booklet, "A Guide for Authors of Memoirs."**
> **Write to Editorial Offices, American Mathematical Society, P.O. Box 6248,**
> **Providence, R.I. 02940.**

MEMOIRS are printed by photo-offset from camera copy fully prepared by the author. This means that the finished book will look exactly like the copy submitted. Thus the author will want to use a good quality typewriter with a new, medium-inked black ribbon, and submit clean copy on the appropriate model paper.

Model Paper, provided at no cost by the AMS, is paper marked with blue lines that confine the copy to the appropriate size.

Special Characters may be filled in carefully freehand, using dense black ink, or **INSTANT** ("rub-on") **LETTERING** may be used. These may be available at a local art supply store.

Diagrams may be drawn in black ink either directly on the model sheet, or on a separate sheet and pasted with rubber cement into spaces left for them in the text. Ballpoint pen is not acceptable.

Page Headings (Running Heads) should be centered in CAPITAL LETTERS (preferably), at the top of the page — just above the blue line and touching it.

LEFT-hand, EVEN-numbered pages should be headed with the AUTHOR'S NAME;

RIGHT-hand, ODD-numbered pages should be headed with the TITLE of the paper (in shortened form if necessary).

Exceptions: PAGE 1 and any other page that carries a display title require NO RUNNING HEADS.

Page Numbers should be at the top of the page, on the same line with the running heads.

LEFT-hand, EVEN numbers — flush with left margin;

RIGHT-hand, ODD numbers — flush with right margin.

Exceptions: PAGE 1 and any other page that carries a display title should have page number, centered below the text, on blue line provided.

FRONT MATTER PAGES should be numbered with Roman numerals (lower case), positioned below text in same manner as described above.

MEMOIRS FORMAT

> **It is suggested that the material be arranged in pages as indicated below.**
> **Note: Starred items (*) are requirements of publication.**

Front Matter (first pages in book, preceding main body of text).

Page i — *Title, *Author's name.

Page iii — Table of contents.

Page iv — *Abstract (at least 1 sentence and at most 300 words).

Key words and phrases, if desired. (A list which covers the content of the paper adequately enough to be useful for an information retrieval system.)

*1991 Mathematics Subject Classification. This classification represents the primary and secondary subjects of the paper, and the scheme can be found in Annual Subject Indexes of MATHEMATICAL REVIEWS beginnning in 1990.

Page 1 — Preface, introduction, or any other matter not belonging in body of text.

Footnotes: *Received by the editor date.
Support information — grants, credits, etc.

First Page Following Introduction – Chapter Title (dropped 1 inch from top line, and centered). Beginning of Text.

Last Page (at bottom) – Author's affiliation.